When Economies Change Paths

Models of Transition in China, the Central Asian Republics, Myanmar & the Nations of Former Indochine Française

When Economies Change Paths

Models of Transition in China, the Central Asian Republics, Myanmar & the Nations of Former Indochine Française

Leo Paul DANA
BA, MBA, PhD

Senior Advisor
World Association for Small & Medium Enterprises, Canada

University of Canterbury, Christchurch, New Zealand

World Scientific
New Jersey • London • Singapore • Hong Kong

Published by

World Scientific Publishing Co. Pte. Ltd.

P O Box 128, Farrer Road, Singapore 912805

USA office: Suite 1B, 1060 Main Street, River Edge, NJ 07661

UK office: 57 Shelton Street, Covent Garden, London WC2H 9HE

British Library Cataloguing-in-Publication Data
A catalogue record for this book is available from the British Library.

WHEN ECONOMIES CHANGE PATHS: Models of Transition in China, the Central Asian Republics, Myanmar & the Nations of Former Indochine Française

ISBN 981-02-4949-7

This book is printed on acid-free paper.

Printed by FuIsland Offset Printing (S) Pte Ltd, Singapore

Dedicated to Ima, Perto, & Teresa

Foreword

In *The Silk Road*[1], L. Boulnois showed how ignorant the Western world has been of the Orient. Not long ago, that may not have mattered. In our global economy, it does, and in this book, Leo Paul Dana takes the reader along the Silk Road and across the transitional economies of the Orient, revealing powerful contrasts within this vastly heterogeneous region.

Internationalisation is no longer an option. Firms must either source internationally or compete in global markets, and many opportunities lie in Asia's transitional economies. Yet, firms considering expansion into transitional economies face difficulties in getting access to reliable information. One can pick up a book about Myanmar and another about Vietnam, but these are seldom comparable. *When Economies Change Paths* is a unique volume, providing readers with a comprehensive introduction to the current situation existing when nations embark on any of several models of transition. Each of the relevant countries has been researched by the same author, using the identical methodology. Every chapter has been sent to various experts for verification. The respectful treatment of different cultures is typical of the author's unbiased approach. For the advanced researcher, this book provides useful citations of over 250 references.

Leo Paul Dana is one of the absolutely finest researchers I have ever encountered. The material he produces is always insightful and of direct practical value. He sees deeply into the cultures of Asia, and he communicates his ideas clearly. The opportunity to learn from his penetrating insights into the transitional economies of this dynamic region can be highly rewarding, both to the business leader and to the general public.

Richard W. Wright, PhD
The E. Claiborne Robins Distinguished University Chair
University of Richmond, Virginia
United States of America

---

[1] Translated by Dennis Chamberlain, London: George Allen & Unwin, 1966.

Contents

Preface

My intrigue with the centrally planned economies of Asia led me to first visit mainland China in 1984, and to explore Burma in 1986. Overwhelmed, I decided to return to that part of the world, for an extended period, to learn more about central planning and about transition. By 1992, I was based in Vietnam, representing a Canadian company.

In 1997, with Professor Richard W. Wright, I took a class of undergraduate and MBA students from McGill University to Vietnam; this was the first time McGill offered Faculty of Management courses in Asia. Shortly thereafter, I served as Visiting Professor on the University of Pittsburgh Semester-at-Sea programme, bringing several hundred North American students to China and Vietnam. My subsequent appointment was at Nanyang Business School, where I was Deputy Director of the International Business MBA Programme; it was there that I began planning the research for this book.

Explicit knowledge is easily identifiable, easy to capture, articulate, and share; implicit knowledge is less easy to transfer. In the transitional economies of Asia, much knowledge is implicit, and the process of obtaining information can be challenging. The attempt to find reliable secondary sources is often futile. Much of that which has been published by government sources appears to be based more on exaggerated figures and propaganda than on fact. I consequently opted to focus my efforts on primary research. Recognising that an etic research design would risk underestimating the impact of culture, I used an emic research design. Inspired by the methods described by Denzin (1978), Morgan (1983), Pasquero (1988), and Patton (1982; 1987), this book is the result of my ethnographic field research.

Methodology included participant observation as well as roundtable discussions and in-depth interviews with entrepreneurs, foreign investors, government officials, policy-makers, and representatives of assistance organisations, including the US Peace Corps and the European Union's Technical Assistance to the Commonwealth of Independent States (TACIS).

Triangulation was used to verify for bias caused by social desirability. This is discussed, in detail, by Arnold & Feldman (1981), Arnold, Feldman & Purbhoo (1985), Crowne & Marlowe (1960), Golembiewski & Munzenrider (1975), Rosenkrantz, Luthans & Hennessey (1983), Thomas & Kilmann (1975), and Zerbe & Paulhus (1987).

While interviewees were usually co-operative, there were unexpected difficulties – cultural, political, physical and logistic. In the steppes of Central Asia, for instance, visitors must drink fermented mare's milk, in order to establish a culturally appropriate setting for discussion. Although the beverage tastes delightful, its alcohol content – coupled with an altitude of 3,000 metres – can detract from one's agenda! In Kazakhstan, I was detained when police suspected my assistant of espionage. In the Kyrgyz Republic, authorities accused us of asking "too many" questions.

In Myanmar, a customs official asked me whether I had any cigarettes. When I said "No," he responded with the question, "Why not?" He then got angry when I explained that I did not smoke. Frustrated, he exclaimed, "Not for you! I need cigarettes for me." In Tajikistan, many people often consume noumia – a drug made from a mixture of asphalt and mineral oil. It is virtually impossible to interview people who are under the influence of this drug. Also challenging was the effect of extreme temperatures. In Turpan, at an elevation of 154 metres below sea level, temperatures can exceed 50°C. In the Taklimakan Desert, the temperature of the sand reaches 65°C. In Tajikistan, cholera, malaria, and food shortages are common. I fell victim to malaria. Transportation was often problematic. In Central Asia, many roads are often blocked – due to snow. In some places, petrol pumps dried up. I relied on fuel trucked in from Uzbekistan. In Uzbekistan, where fuel is abundant, everything came to a halt when my driver demanded an exorbitant sum of cash, threatening to strand us.

I should also mention difficulties related to inter-ethnic violence, in Tajikistan. Even the American embassy in Dushanbe operates only when security allows. Although I followed directives from the CIA and took extreme caution, incidences resulted in evacuation to Hong Kong.

I was, nevertheless, determined to put together this book about transition in Asia. After many months of struggle, I am pleased to share with you the first book of its kind. Enjoy it!

Leo Paul Dana, BA, MBA, PhD

The great number of men who sail the seas do so for love of gain, not love of knowledge, and never dream, in their blindness and greed, that navigation itself would become safer with an increase in knowledge.

– Pliny the Elder,
in ***Natural History***

Acknowledgements

The research leading to this book was conducted on location, in the transitional economies of Asia. For travel to Asia, the author received funding from McGill University, from the University of Oxford, and from the University of Pittsburgh.

In the process of preparing the manuscript, each chapter was sent to experts for review. The following kindly volunteered their time, providing inspiration, encouragement and constructive comments:

- Lola ASKAROVA, Business Consultant, Almaty, Kazakhstan
- Lady BORTON, Field Director, American Friends Service Committee, Hanoi, author of *After Sorrow*
- Professor David P. CHANDLER, Professor Emeritus of History, Monash University, Melbourne, Australia; Adjunct Professor, Johns Hopkins University, Washington, DC; Adjunct Professor, Georgetown University, Washington, DC; and Visiting Professor, Cornell University, Ithaca, New York
- Professor CHEW Soon Beng, Professor of Economics, Director of the Asian Commerce and Economics Studies Centre, Nanyang Technological University, Singapore
- Teresa E. DANA, Department of Management, University of Canterbury, Christchurch, New Zealand
- Sophal EAR, formerly at the World Bank and currently at the University of California at Berkeley, Berkeley, California
- Professor Hamid ETEMAD, former Associate Dean & Director of the McGill China Programme, McGill University, Montreal, Canada
- Professor Robert HAMILTON, Dean of Commerce, University of Canterbury, Christchurch, New Zealand
- Professor Jan JOHANSON, Professor of International Business, Centre of International Business Studies, University of Uppsala, Uppsala, Sweden

- Professor Raymond KAO, Founder & Editor, *Journal of Small Business & Entrepreneurship*, University of Toronto, Toronto, Canada
- Professor Jerome A. KATZ, Mary Louise Endowed Professor of Management, Saint Louis University, St. Louis, Missouri
- Professor W. F. Fred KIESNER, Conrad Hilton Chair of Entrepreneurship, Loyola Marymount University, Los Angeles, California
- Winn KYI, General Manager, Myanma Export Import Corporation, Merchant Street, Yangon, Myanmar
- Professor Philippe LASSERRE, Professor of Strategy & Asian Business, Starr Foundation Research Fellow, Coordinator, Asian Business Area, INSEAD, Fontainebleau, France
- Professor Nancy K. NAPIER, Professor of Business and Economics, Boise State University, Boise, Idaho
- Professor Robert PEFFERS, former Dean, University of Maryland University College, Asian Division
- Professor Michael PENG, Ohio State University, Columbus, Ohio
- James PICKFORD, Executive Editor, *Financial Times*, London, England
- Sheridan PRASSO, *BusinessWeek*
- Dr. Herbert Leo SCHARF, Harvard University, Cambridge, Massachusetts
- Professor David I. STEINBERG, Director of Asian Studies, School of Foreign Service, Georgetown University, Washington DC
- Professor Naranhkiri TITH, Professor of International Economics and Asian Studies, John Hopkins University, Washington DC
- Myoe WIN, Director of the Ministry of Commerce Directorate of Trade, Yangon, Myanmar
- Professor Richard W. WRIGHT, The E. Claiborne Robins Distinguished University Chair, University of Richmond, Richmond, Virginia
- Professor Joseph Jermiah ZASLOFF, Professor of Political Science, University of Pittsburgh, Pittsburgh, Pennsylvania
- Professor Tela ZASLOFF, Semester-at-Sea Program, University of Pittsburgh, Pittsburgh, Pennsylvania

Chapter 1

Introduction

Toward Market Economies

Whether transition is taking place gradually or rapidly, alongside political reform or in its absence, the mindset of people has often held onto perceptions of former times. Consequently, in order to gain an understanding of the behaviour of entrepreneurs and of the nature of their enterprises, one must first become familiar with a variety of explanatory variables, including culture, history, and government policy. Transition is a function of all of these causal factors; furthermore, transition takes on a particular accent where different ethnic groups meet. This book examines transition in the context of cultural influences, historical experiences, ethnicity, demographic shifts, and pluralism.

Interest in transitional economies has been on the rise and much has been written about Eastern Europe after the fall of the Iron Curtain. Examples include research conducted in Albania (Dana, 1996a; 2000b), Bulgaria (Dana, 1999a), Croatia (Franicevic, 1999; Martin & Grbac, 1998), the Czech and Slovak Republic (Rondinelli, 1991), the Czech Republic (Dana, 2000d; Sacks, 1993), Estonia (Liuhto, 1996), the Former Yugoslav Republic of Macedonia – FYROM (Dana, 1998c), Hungary (Hisrich & Fulop, 1995; Hisrich & Vecsenyi, 1990; Noar, 1985), Latvia (Peng, 2000), Moldova (Dana, 1997c), Poland (Arendarski, Mroczkowski & Sood, 1994; Sachs, 1993; Zapalska, 1997), Russia (Ahmed, Robinson & Dana, 2001; Bruton, 1998; Hisrich & Gratchev, 1993; Robinson, Ahmed, Dana, Latfullin & Smirnova, 2001), Slovakia (Dana, 2000d; Ivy, 1996), and the Ukraine (Ahmed, Dana, Anwar & Biedyuk, 1998).

Researchers have also examined the economies of former Soviet allies overseas. This includes Angola (Gray & Allison, 1997), Cuba (Dana, 1996c), and Mozambique (Dana, 1996b).

Still a Hero

Others have researched transition in Asia. Among the numerous accounts of transition in China are Beamish (1993), Chau (1995), Chow & Tsang (1995), Dana (1998b; 1999b; 1999c), Dandridge & Flynn (1988), Fan, Chen & Kirby (1996), Lombardo (1995), Overholt (1993), Peng (2000), Shirk (1993), Siu & Kirby (1995), Wei (2001), and Williams & Li (1993). Brown & Zasloff (1999) and Dana (1999c) have provided renditions of transition in Cambodia. Dana (1995b; 1999c) focused on Laos. Dana (1994a; 1994b; 1994e; 1999c), Peng (2000) and Tan & Lim (1993) reported on Vietnam. While there have been countless country-specific publications on transition from centrally planned to market economies, the purpose of this book is to provide a one-volume overview of transition in Asia.

Models of Transition

Several models of transition have been identified in the literature. These include the Yugopluralist Model; the Model of Restructuring (*Perestroika*) by Decree; the *Perseritje* Model; and Models of Gradual Transition, including the *Doi-Moi* (Renovation) Model. These shall now be discussed.

The Yugopluralist Model

In Yugoslavia, transition was initiated at the local level. This resulted in an economic model unique to Yugoslavia, and a function of its pluralism – the Yugopluralist Model (Dana, 1994f). This model was a result of a weak central government, coupled with the cultural heterogeneity of its constituent republics. As members of the federation gained substantial autonomy, each one embarked on a different programme of transition.

Moving at Different Speeds in Yugoslavia

Restructuring (*Perestroika*) by Decree

In contrast to transition in the Yugoslav federation, *perestroika* entails change dictated by a central government. Involving transition by decree (Dana, 1994d; 1998a), its objective is to completely restructure an economy, as implied by the Russian word *perestroika* – restructuring. Although Russia had neither a long history of capitalism, nor a culture that traditionally valued entrepreneurship, *perestroika* legislation abruptly decreed a change in the economic system of the state, phasing out communism in favour of capitalism. It was felt that the expedient liberalisation of prices and privatisation of state firms would lead to rapid transition. Variants of this model have met different levels of success across the Union of Soviet Socialist Republics (USSR) and among member countries of the Council for Mutual Assistance (COMECON).

Typical Queues in the USSR

A drawback of rapid transition by decree is that it can destabilise a country, generating unemployment and social problems. When, in 1990, the German Democratic Republic (East Germany) rushed into transition, its unemployment skyrocketed immediately from almost nil in June, to 200,000 following economic unification in July and possibly over 2,000,000 after political unification in October (Dana, 1994c).

The Last Goose-step in East Berlin

German Watchtower, No Longer in Use

Former Boundary Between East Germany and the West

Abandoned Border Post

Rapid transition can be a shock to society, and this approach has been referred to as "shock policy" (Dana, 2000a). President Karimov of Uzbekistan used the term "shock therapy." Peng (2000) described this as the "big bang approach." As will be discussed in Chapter 9, upon gaining its independence, Uzbekistan established policies that were clearly opposed to such shock therapy; it opted instead for gradual transition.

The *Perseritje* Model

Reform in Albania was even more radical than that in the USSR and the COMECON countries. In 1986, there were only forty automobiles in Albania; there were neither bananas nor Coca-Cola in this country where imports were few. Yet, once the nation moved away from Enver Hoxha's ideology and embarked on its *Perseritje* Model of transition, the Bretton Woods institutions considered Albania to be one of the most successful transitional economies in Europe. The *Perseritje* Model is analysed in detail, in Dana (1996a; 2000b).

Using Enver's Book to Wrap Sausages

Perseritje **Legalised Bananas in Albania**

Models of Gradual Transition

While governments in Eastern Europe formally denounced communism and officially embraced democracy, free press and the values of a free-market economy, transition in Asia involves models of its own. In contrast to the big bang approach – which proclaimed immediate transition to capitalism – China implemented a model of gradual transition, tolerating private enterprise as a complement to the centrally planned state sector, but not as a replacement. Kruft & Sofrova (1997) emphasised the gradualism.

This concept of harmony – between government firms still operating under a system of central planning, and private enterprise being a complement to the state sector – is also characteristic of Vietnam's *Doi-Moi* Model. The Vietnamese word *doi-moi* literally means "renovation." No major political reform is implied here; instead, this model is compatible with restoring the prestige of the ruling Communist Party of Vietnam. The government affirmed its commitment to free enterprise within the context of socialism.

Economic Sectors

Where governments have clung on to socialist ideology, there tends to be a large sector of the economy that is state-controlled, and it operates along side the traditional bazaar and the more modern firm-type sector. It is useful, therefore, to distinguish among these very distinct sectors of economic activity, which co-exist in the transitional economies of Asia. The bazaar, the state-controlled planned sector, and the firm-type sector are components of the formal economy, some of the features of which are summarised in Table 1.1.

In addition, the parallel economy includes informal economic activity; internal economic activity with no transaction; covert economic activity; and fictitious economic activity. Discussions of these shall follow.

THE BAZAAR	STATE-CONTROLLED PLANNED SECTOR	FIRM-TYPE SECTOR
Focus on personal relationships	Focus on bureaucracy	Focus on impersonal transactions
Segmentation refers to producers	Segmentation not considered	Segmentation refers to the market
Competition refers to tension between buyer and seller	Competition is deemed unnecessary, as the state declares a monopoly	Competition is an activity which takes place among sellers
Prices are negotiated	Prices are dictated by the state	Prices are indicated by the vendor

Table 1.1. Sectors of the Formal Economy

The Bazaar

The bazaar is a social and cultural system, a way of life and a general mode of commercial activity, which has been in existence for millennia. In the bazaar, economic transactions are *not* the focus of activities; instead, the focus is on personal relationships. In this scenario, consumers do not necessarily seek the lowest price or the best quality. An individual gives business to another with whom a relationship has been established, to ensure that this person will reciprocate. *Reciprocal preferential treatment reduces transaction costs.* The multiplicity of small-scale transactions, in the bazaar, results in a fractionalisation of risks and therefore of profit margins; the complex balance of credit relationships is carefully managed, as described by Geertz (1963).

Prices in the bazaar are negotiated, as opposed to being specified by the seller. In contrast to the firm-type sector, in which the primary competitive stress is between sellers, the sliding price system of the bazaar results in the primary competitive stress being between buyer and seller (Parsons & Smelzer, 1956). The lack of information results in an imperfect market and with few exceptions, such as basic food staples, retail prices are not indicated; rather, these are determined by negotiations. The customer tests price levels informally, before bargaining begins. It is often the buyer who proposes a price, which is eventually raised. As discussed by Geertz, the "relatively high percentage of wholesale transactions (*i.e.*, transactions in which goods are bought with the express intention to resell them) means that in most cases both buyer and seller are professional traders and the contest is one between experts (1963, p. 33)."

Once a mutually satisfactory transaction has taken place, the establishment of a long-term relationship makes future purchases more pleasurable, and profitable. As noted by Webster (1992), building long-term relationships can be viewed as a social and economic process. Unlike Western relationship marketing, which is customer-centred, whereby a seller seeks long-term business relationships with clients (Evans & Laskin, 1994; Zineldin, 1998), the focus in the bazaar is on the relationship itself. In the bazaar, *both* the buyer and the seller seek a personal relationship.

Firms in the bazaar are not perceived as rivals of one another. There is minimal – if any – brand differentiation among merchants. Vendors do not necessarily seek to optimise monetary gain. Economic rationality is not an issue.

In contrast to the Occident, where segmentation refers to the market, in the bazaar economy, segmentation refers to the clustering of producers and retailers; street-names reflect this. In Yangon – formerly Rangoon – the Chamber of Commerce and Industry is located on Merchant Street. The street-map of Hanoi includes Broiled Fish Street, Jewellers Street, Sweet Potato Street, and Tin Street. Long gave an account of his observations in Hanoi, "Street names in the teeming native-Chinese section are a guide to the shopper. Each bears the name of the product traditionally sold there – silk, tin, scales, spice, brass, paper, jewelry (1952, pp. 315-316)."

Bazaar in Hoi An, Vietnam

Likewise, Passantino described the bazaar in Kunming, China, "There is also a street of beggars...There are streets for pig auctions and for prostitutes; there is one for banks...(1946, p.142)." Even today, in the bazaar economy, shops still are clustered according to the goods offered therein.

When examining the bazaar, it is important to understand the relationships among the players within it, their organisation and their economic principles. The bazaar is a hub of information exchange. Buyers and sellers express intentions, and an intricate network of relationships facilitates transactions. As explained by Christian, "News spreads rapidly in a Shan States bazaar. Within the hour I was offered a half a dozen old pistols (1943, p. 504)."

While the entrepreneur described by Schumpeter (1912; 1928; 1934; 1939; 1942; 1947; 1949) is an innovator who causes disequilibrium to profit therefrom, the entrepreneur of the bazaar may simply identify an opportunity for profit – rather than create one. In this way, the entrepreneur of the bazaar corresponds to that of the Austrian school (Kirzner, 1973; 1979; 1982; 1985). Geertz (1963) and Dana (2000b) discuss, in detail, phenomena of the bazaar.

No Brand Differentiation at the Bazaar

The State-Controlled Planned Sector

In transitional economies, state firms are remnants of the communist model – a doctrine first published in German (Marx & Engels, 1848), in Russian in 1882, and in English in 1888. This model assumed that a central office was in the best position to balance supply and demand. The focus of the state-controlled planned sector is thus neither on transactions nor on relationships, but rather on the state bureaucracy.

When the state produces everything, centralisation rules out competitors. Barriers to trade, coupled with an import-substitution policy, ensure that competition is not a factor. Since demand exceeds supply, marketing is not necessary and segmentation need not be considered. Prices are a function of the government's bureaucracy. Dalgic (1998) reported on an empirical study, which found that state-owned firms had much less of a market orientation, than did private companies.

State Bakery

The Firm-Type Sector

The firm-type sector involves a mode of commercial activity such that industry and trade take place primarily within a set of impersonally defined institutions, grouping people according to organisation and specialisation. It is assumed that profit-maximising transactions will occur based on rational decision-making, rather than based on the nature of personal relationships. The focus is on impersonal transactions. Weber's (1924) thesis applies here.

In this sector of the economy, the decision space is occupied by product attributes; the buyer and seller are secondary, if not trivial, to the transaction decision. The interaction between the buyer and the product is deemed more important than that between the buyer and the seller. Transactions are based on economic rationality and are therefore impersonal in nature. Competition is an activity that takes place between sellers, who engage in segmentation, in order to partition the market into like-groups of predictable consumers. Prices are tagged, reflecting market forces.

While Western marketing principles (Gronroos, 1989) apply to this sector, market-orientation is linked to the maturity of the industrialisation process (Seglin, 1990). Where industrial development is limited, the framework for economic transition may have to rely more on new ventures and on joint initiatives.

Firm-Type Sector in St. Louis, Missouri

The Parallel Economy

Under central planning, the lack of a legal market economy led to permanent shortages. Survival strategies often involved the emergence of entrepreneurs in the parallel economy, where inefficient regulations could be circumvented. According to Grossman (1977), this underground activity increased the overall efficiency of resource allocation under central planning. The problem is that a mindset evolved, equating efficiency with the evasion of regulation.

Recent years have been characterised by economic reform, regulatory reform, and change in the mindset of people. However, change in mindset has not kept up with changes in regulatory framework (North, 1990). Since these have not been evolving at the same pace, new problems have become associated with transition. As a consequence of their experience under central planning, people came to equate entrepreneurship with the avoidance of communist law. When new regulations were introduced to usher in market economics, people continued to circumvent business law. As noted by Feige & Ott (1999), during transition, evasion and non-compliance with new rules renders them ineffective. Thus, where economic reform has been faster than the ability of people to adapt, inertia has delayed actual transition. Štulhofer (1999) used the term "*cultura inertia*" to describe a collectivist legacy that has survived from the past. Especially among the elderly, there is still a distrust of the state, of banks[2] and of legal institutions. Conditions in transitional economies thus make the parallel sector very popular, avoiding all forms of taxation. In transitional economies that lack developed market institutions, it is common to have a high proportion of underground activities. O'Driscoll, Holmes & Kirkpatrick (2001) reported a black market in Laos, larger than the formal economy.

The popularity of the parallel economy is no surprise, considering the low initial role of legitimate private enterprise, coupled with a high degree of liberalisation, and hindered by the lack of macro-stability in the absence of a sufficiently developed legal framework. As illustrated in Table 1.2, the parallel economy may be informal; internal; covert; or fictitious.

[2] When the author and his assistant were robbed by a bank employee, at a bank in Kazakhstan, several people further confirmed the need to avoid banks.

CATEGORY	EXAMPLES
Informal Economic Activity (with business transaction)	Barter, street vending, unrecorded cash sales
Internal Economic Activity (with no business transaction)	Subsistence agriculture, hunting, fishing
Covert Economic Activity (with illegal business transaction)	Prostitution, smuggling, trade in illegal drugs
Fictitious Economic Activity	Foreign devil company

Table 1.2. Activities of the Parallel Economy

Informal Economic Activity

Barter, selling from an impromptu stall and itinerant vending are considered forms of informal economic activity. Unrecorded cash sales circumvent taxation as well as regulation.

In transitional economies, private enterprise is often concentrated in the informal sector. This is illustrated by the large numbers of self-employed vendors in the profitable distribution of goods and services. The law is often bent, but authorities generally tolerate the sector. A relevant discussion from Dana (1992) is presented concisely by Chamard & Christie (1996).

For many entrepreneurs in some transitional economies, legal transactions are often limited to informal barter. The transition from such a traditional system to a modern cash economy will require cognitive innovation. A prerequisite will be for people to internalise new notions of measurability, to encourage the use of cash. New elements of formality, and impersonal structure, will have to be introduced and internalised, before new institutions can function effectively.

Internal Economic Activity

This category of economic activity is best described as internal, because *no business transaction* takes place. Wealth is created, but nothing is sold for profit. That which is created is consumed or saved for personal use. In transitional economies, internal subsistence activity is often necessary, as a means to adapt to rapid reform.

Examples of internal economic activity include subsistence agriculture, and subsistence fishing. Both are legal, but involve no market transaction external to the producer. These are, therefore, forms of internal economic activity.

While internal economic activity exists – as an activity of choice – even amidst the most advanced and industrialised backdrop (Dana, 1995a), for some people in transitional economies, this is the only strategy for survival.

Toiling for Survival

Covert Economic Activity

This category of economic activity involves business transactions, which are illegal, and therefore conducted in a covert way, in order to avoid punitive measures from law-enforcing authorities (Haskell & Yablonsky, 1974; Henry, 1978). Prostitution, which Cantillon (1755) considered a form of entrepreneurship, falls in this category.

In Cambodia, illegal logging is an economically significant covert activity. In the Kyrgyz Republic and in Myanmar, smuggling has been on the rise (O'Driscoll, Holmes & Kirkpatrick, 2001).

Covert activity promises fast cash. Reporting on the extremely high level of underground activity in Kazakhstan, Glinkina (1999) explained that according to expert forecasts, heroin technology was likely to grow in Tajikistan and in Turkmenistan.

Fictitious Economic Activity

Fictitious economic activity has been created to facilitate circumvention of the law; this "implies speculative transactions and different kinds of swindles with a view to receiving and transferring money, including contrived rent-seeking (Glinkina, 1999, p. 102)." As shall be shown in Chapter 14, much of this has been taking place in Vietnam, where "foreign devil" companies have been used to set up fictitious economic activity.

Melting Pot Pluralism & Structural Pluralism

Where groups with unlike spheres of values co-exist, the result is a pluralistic society. Norwegian anthropologist Frederik Barth[3] placed great emphasis on the existence of different spheres of values. Central to his discussion is the concept of the entrepreneur as being an essential broker, mediating boundary transfers in this situation of contacts between cultures.

[3] See also: Barth (1963; 1966; 1967a; 1967b; 1981).

By being active in the transformation of a community, entrepreneurs are the social agents of change.

Enterprise is very much influenced by the nature of pluralism. It is, therefore, important to distinguish between **melting pot pluralism**, and **structural pluralism.** When people, from different cultures, share activities in a secular mainstream arena, the expression of cultural differences tends to be limited to private life. Often, employment is shared in a common sphere of life, while cuisine, customs, languages and religion are a domestic concern. This form of socio-economic pluralism is referred to as **melting pot pluralism**, and this is descriptive of the situation in the United States.

In contrast, **structural pluralism** involves a society with different cultures that do *not* share a secular mainstream arena. In such a case, there is minimal interaction across cultures. Rather, each ethnic group has its distinct institutions, and members of a given community have a lifestyle that is incompatible with that of people from other backgrounds. This type of pluralism is prevalent in Xinjiang (China), and in the Central Asian republics.

Toward an Understanding of Transition

Although the West has provided much funding to transitional economies, and infrastructure has been greatly improved in recent years, complaints are often heard about the problems arising from transition. Outsiders often fail to realise that transition to a market economy requires more than funding and infrastructure. Transition also involves mindset. Business takes place between people, and the interaction between the parties does not take place in a vacuum, but rather it is part of a social system, as discussed by Hakansson (1982).

Central to transactions are the cultural assumptions of a social system. In the West, these are implicit because it is assumed that everyone knows about them; marketing takes place in the context of a firm-type economy. In transitional economies, environmental factors must not be ignored. As discussed by Huntington (1996), globalisation has not led to a single world culture.

The move to a market economy – transition – is process-driven, and this necessitates the understanding of people and their culture. The focus of the following chapter, therefore, is culture. The subsequent chapters focus on different areas of Asia. Throughout this book, the symbols $ and ¢ refer to the currency of the United States.

Chapter 2

Culture & Enterprise[4]

Introduction

Globalisation may have given rise to a global village, but has not produced a homogeneous world culture. This is important, because culture shapes managerial assumptions. Formerly closed economies are opening up, but they are often misunderstood. While that which goes on around the negotiation table is important, everything that surrounds a deal is equally influential – spoken or unspoken.

Business increasingly involves interaction with foreign suppliers, employees, distributors, clients and government officials. Each operates in a particular environment, where interactions involve implicit and explicit assumptions. This is a function of the complex belief systems, cultural values and attitudes that dictate accepted norms of conduct.

Many managers report frustrations when doing business in Asia, particularly when delaying tactics appear during negotiations. Du Ding-Ping, in his book about how to do business with Americans, explains that delays help the Chinese during negotiations with Westerners. Perhaps knowing this may be a good reminder of the Western notion that "patience is a virtue" and that it might be beneficial to adhere to this virtue when doing business with Asians. Developing the ability to be patient and tolerant can surprisingly speed up negotiations; inflexibility and haste may, more often than not, lead to a bad relationship, what the Chinese call *guanxi gao jiang*.

---

[4] This chapter draws heavily on material published in an article commissioned by the *Financial Times* (Dana, 2000c).

It is essential for people in business to understand cultural differences. Yet, many Westerners often fail to acknowledge the extent to which culture affects enterprise.

It is not a matter of noting the "different ways" that "those" people have; it is considerably more useful to understand the rationale behind these ways. Everything has a reason, and actions may have implications. In contrast to the short-term focus of many companies in the West, Asians often have a longer-range plan. Also, Asian entrepreneurs are generally reluctant to share their implicit assumptions with outsiders.

Many Asians have mastered the comprehension of Western culture and have developed the ability to adapt to it; they have built on the knowledge conveyed to them by a variety of sources. In contrast, Westerners seldom have the patience to master an Asian language, let alone one of its cultures. Complex cultural environments require fluency not only in words, but also more importantly in understanding of what is not said – and this takes time, patience and dedication to learn. Westerners may argue that time is money and they cannot afford to learn everything it takes before even discussing a deal. In the long term, however, more knowledge does pay off.

The Chinese Experience

The dominant culture in Asian business is that of the Chinese. This is not surprising, considering that ethnic Chinese are well represented in the realm of business around the world and very concentrated in Asian countries. In Ho Chi Minh City – formerly Saigon – for instance, 12% of the population is ethnic Chinese, and this group controls half of the local economy.

Given the widespread influence of Chinese culture, one would think that it is appropriate for Western businesses to take the time to learn something about this culture and how it affects business; yet, even large and global players have made costly errors. In 1979, Northern Telecom was approached to provide new switching equipment for China. The communications giant underestimated the importance of "face" and embarrassed Chinese officials. When the company tried to get a deal with China in 1987, the Chinese officials remembered that Northern Telecom had "walked away from negotiations in 1979." This time, the tables were turned: although Northern Telecom was eager to do business, the Chinese negotiators were not.

Networking as Viewed from the Occident

Networking involves calling upon a web of contacts for information, support and assistance. Aldrich & Zimmer (1986) integrated social network theory into the study of entrepreneurship. They linked entrepreneurship to social networks. Carsrud, Gaglio & Olm (1986) also found networks important to the understanding of new venture development. Aldrich, Rosen & Woodward (1987) studied the impact of social networks on profit, as well as on business creation; they found network accessibility significant in predicting new venture creation. Likewise, Dubini & Aldrich (1991) found networks central to entrepreneurship. Gomes-Casseres (1997) focused on the alliance strategies of small firms. Anderson (1995) and Johanson & Associates (1994) studied the effect of business networks on internationalisation of firms. Indeed, networks can influence a firm's degree of internationalisation (Dana, Etemad & Wright, 2000).

Analysing Asian entrepreneurs in Britain, Aldrich, Jones & McEvoy found that their sample benefited from "certain advantages denied non-ethnic competitors (1984, p.193)." They found a strong internal solidarity in the ethnic enclave. Given that the possibility of exploiting opportunities appears to be linked to the internal organising capacity of a group – such as creating an ethnic network – Auster and Aldrich (1984) concluded that the ethnic enclave reduces the vulnerability of small firms, by providing an ethnic market and also general social and economic support, including credit. In a study of Asians in Atlanta, Min & Jaret (1985) found networks of co-ethnics to be a source of manpower for entrepreneurs. Boissevain and Grotenbreg (1987), in their study of the Surinamese in Amsterdam, suggested that access to an ethnic network is an important resource for entrepreneurs; they noted, for instance, that networks could provide introductions to wholesalers and warnings of government inspections. Others who focused on ethnic enterprise include Aldrich & Waldinger (1990), Cummings (1980), Dana (1995a), Jenkins (1984), Light (1972; 1984), Light & Bonacich (1988), Min (1984; 1986-7; 1987), Portes & Bach (1985), Portes & Jensen (1987; 1989; 1992), Sanders & Nee (1987), Waldinger (1984; 1986a; 1986b), Waldinger & Aldrich (1990), Waldinger, Aldrich & Ward (1990), Waldinger, McEvoy & Aldrich (1990), Ward (1987), Ward & Jenkins (1984), Wong (1987), and Wu (1983).

Networking in Transitional Economies

Aspects of networking reflect historical, cultural and demographic factors. In contrast to the situation prevailing in many industrialised countries, where voluntary migration prompted members of some immigrant minorities to create networks to help other members of their community (Acs & Dana, 2001; Boissevain and Grotenbreg, 1987; Brenner & Toulouse, 1990; Dyer & Ross, 2000; Iyer & Shapiro, 1999; Light, 1984), much of the demographic shifting in transitional economies resulted from forced migration.

Immigration to Central Asia has often been the result of civil engineering by the Soviets – rather than a voluntary migration of entrepreneurs with a high need for achievement (McClelland, 1961). In Central Asia and in Indochina, the situation has been such that there has been considerable emigration, as opposed to immigration. Despite the absence of immigrant networks in these environments, there are nevertheless ethnic networks and these take on a distinct flavour. The Muslims of Central Asia, for instance, have developed networks of like-minded co-ethnics in neighbouring countries, and this has enhanced opportunities for trade. International transactions are facilitated by the fact that members of ethnic networks speak the same language, and share similar values, influenced by the same religion. Yet, even where religion is shared, there can be important differences in language. The Kazakh language is a Turkic one; the Tajiks, in contrast, speak a variant of Farsi.

In spite of the political divide, which separated the Kazakhs of China from those in the Soviet Union, the Kazakhs in China feel that they have more in common with their co-ethnics in the republics of Central Asia, than with the Han-Chinese who rule them. The Han-Chinese read Mandarin and eat dishes that include pork. The Kazakhs – like the Uygurs and other Muslims in China – are often trained in Arabic script, and of course they would feel uncomfortable at a table where pork is being served. Along similar lines, Uzbeks in the Kyrgyz Republic are displeased with the secular attitude of that republic; they feel comfortable doing business with other Uzbeks. Indeed, entrepreneurs are comfortable doing business with like-minded people, whom they understand, and with whom they get along.

Inevitably, entrepreneurs form networks (Dana, 2001). Where civil servants are poorly paid, it is no secret that bribes are welcome. In these scenarios, business networks include government officials, as well as entrepreneurs.

Roles in Society

Confucianism teaches that people are *not* equal. To this day, people in China are defined by their role in society and contribution to it. Status is influenced by relationships and these are attached to implicit duties and obligations. People are viewed as relation-oriented beings, regulated by cardinal relationships that dictate an individual's obligations – *renqing* – toward other people. Observance of proper relationships is essential for the smooth functioning of society.

In Laos, the ethnic Chinese have become highly successful entrepreneurs, partly thanks to the lack of a local entrepreneurial culture. Business activities in this country have not been traditionally associated with high social status. Stemming from religious beliefs, cultural values in Laos emphasise the elimination of desire. Since commerce was traditionally perceived as a means to satisfy desire, social forces discouraged enterprise, and trade has usually been the role of those with inferior social standing. The communist takeover further discouraged entrepreneurial spirit.

In Laos, the culture makes it important to extinguish unsatisfied yearnings. An important doctrine, here, focuses on suffering caused by unsatisfied wishes. Assuming that an unsatisfied craving causes suffering, then suffering can be eliminated if its cause – desire – is suppressed. According to this ideology, a respectable person strives to eliminate a perceived material need, rather than working toward the satisfaction of materialistic wants. Not surprisingly, ethnic minorities, especially the Chinese, dominate private enterprise, in Laos.

Guanxi

In the Chinese realm of business, a special relationship – *guanxi* – involves the exchange of favours, usually involving position or rank. It is similar to insurance, inasmuch as favours are registered – like premiums – so that benefits may be obtained if and when required. *Guanxi*, composed of two words, *guan* – meaning "to close up" – and *xi* – "to tie up" – expresses the notion of being an insider to a relationship network that involves obligations. *La guanxi* refers to the act of getting on one's good side.

Mandarin is Common in Laos, Along With French

Chinese Entrepreneurs in Vientiane

While Western business-people usually have a general understanding of this concept, relatively few Westerners realise that the mutual obligation is a function of one's ability to help, explaining why the weaker party excepts to receive more, in exchange for less. Furthermore, *guanxi* is not easy for Westerners to develop. *Guanxi* is intuitive, rather than rational, and it is not limited to business hours.

In the West, relationship marketing – the pursuit of customer loyalty – involves the fostering of long-term alliances with customers; in a similar fashion, *guanxi* involves building a long-term relationship, based on trust and mutual exchange, in order to secure customer loyalty and good working relations. *Guanxi* forms a bond among buyers and sellers and between suppliers and producers.

Westerners seeking joint venture partners in China should seek the ideal partner using connections maintained through *guanxi*; simultaneously, the foreigner would benefit by linking up with a company that has enough *guanxi* to get things done efficiently in China. For companies regulated by legislation such as the American Foreign Corrupt Practice Act, this is not an easy task; difficulty increases if and when a joint venture is established and the venture must maintain the *guanxi* of counterparts in China. Furthermore, it is important to look at the whole – not simply the parts.

Guanxi has long been important in Chinese business circles. The West has known contract law since the 1700s; the purpose of such legislation is to give business an assurance that deals will be honoured. In the absence of an elaborate contract law, the Chinese have relied on *guanxi* for the same assurance. When a relationship is more valuable than a transaction, then it is likely that the transaction will be smooth. Who would tarnish a relationship for a single transaction? No contract is necessary. The sense of obligation would come from the relationship, rather than a piece of paper.

On the other hand, in the absence of a strong relationship, there exists the possibility that one of the parties might ignore a contract. Although it had 18 years remaining on its lease, McDonald's was evicted from a Beijing site in favour of a newcomer with stronger *guanxi*. In Chinese circles, it is therefore more crucial to monitor a relationship than a transaction. Where a strong relationship exists, problems can always be solved. The long-term benefit of *guanxi* becomes obvious. During the process of cultivating *guanxi* within business relationships, customer loyalty evolves naturally, while bonds are created with suppliers and with creditors. *Guanxi* rests on the moral premise of *renqing*, as a justification for social exchange. In a Chinese cultural setting, *guanxi* is the norm for carrying on business.

Adapting to the Environment

One might argue that Western companies also value relationships and that these hold mutual obligations. There is, however, an important difference between the Chinese and Western views of relationships. *In the West, successful transactions lead to good relationships. In Chinese circles, one builds relationships in order to initiate transactions; the common belief is that if a relationship is built properly, then profitable transactions will follow.*

While *guanxi* is characteristic of enterprise in China, *blat* is the Russian equivalent. This notion has remained important in the Central Asian republics of the Commonwealth of Independent States (CIS), even today. Peng & Heath (1996) discuss the importance of such connections, explaining that reciprocal preferential treatment reduces transaction costs. Williamson (1985; 1996) focuses on the notion of transaction costs.

Fertilised Duck Eggs Are Considered a Delicacy, Raw

Every region has a predominant culture and there are many intricacies intertwined into each one. Just as Europe hosts many different cultures, there is neither one single culture nor one single way of doing business in Asia. The key for those attempting business there is to have patience and to learn as much as possible about what is important to each company, manager and worker.

As stated so well by Lasserre & Schütte, "*The West cannot expect to compete successfully with Asia as long as the Asians know more about the West than the West knows about Asia* (1995, p. xv)." The following chapters provide overviews of the environment for enterprise in China, the Central Asian republics of the CIS, Myanmar and Indochina.

Rice is the Vietnamese Word for Food

Chapter 3

The People's Republic of China (PRC)[5]

Introduction

The People's Republic of China (PRC) – commonly referred to as mainland China, or simply China – covers 9,596,960 square kilometres, neighbouring Afghanistan, Bhutan, India, Kazakhstan, the Kyrgyz Republic, Laos, Mongolia, Myanmar, Nepal, North Korea, Pakistan, Russia, Tajikistan, and Vietnam. China has shores along the East China Sea, the South China Sea, and the Yellow Sea. Unlike the situation prevailing in much of Eastern Europe, in Cambodia and in other states that have abandoned communist ideology in favour of capitalism, entrepreneurship in the PRC was introduced by the government as a supplement to the socialist economy. Despite its success, in June 2001, China's Prime Minister Zhu Rongji declared that he was slowing down transition in the PRC, due to difficulties created by economic reform.

While the *guoying qiye* – literally, state-run enterprise – is the Chinese term to describe a collective enterprise, the *siying qiye* is defined as a private enterprise owned by entrepreneurs and providing employment for eight or more people. Smaller firms, with fewer than eight people are referred to as *getihu*.

---

[5] This chapter includes information obtained from: the China Council for the Promotion of International Trade (CCPIT); the China Individual Labourers Association; the China International Trust Investment Co. (Beijing); the Department of Science and Technology for Rural Development; the Economic Management School of Shanghai University of Technology; the Industrial and Commercial Bureau; Kunming Foreign Economic Relations and Trade Commission; the Ministry of Agriculture (particularly its Department of Township Enterprise); and the Ministry of Foreign Economic Relations and Trade, Market and Trade Development Division. The chapter also includes material that first appeared in Dana (1998b; 1999b; 1999c).

Beijing

Along the Silk Road

Potato Dealer

Zhong Xi Hotel, Shanghai

Historical Overview

China's first dynasty, the Xia, was established over 4,000 years ago. In 221 BC, Emperor Qin Shi Huang of the Qin dynasty (221 BC to 206 BC), united China, centralised his authority over the nation, banned slavery, and initiated the transition to feudalism.

The Hans (206 BC to 220 AD) entrenched the feudal system in China, and allowed the export of silk. During a horse-buying mission in 138 BC, Han-Chinese traders realised that Central Asian merchants were willing to pay a high price for silk, and this soon became China's main export. Agriculture, handicrafts, shipbuilding and weaving were developed during the Han dynasty. Also during this era, the armillary sphere and the seismograph were invented in China. Table 3.1 lists Chinese dynasties.

For many years, international trade occurred not because of encouragement by rulers, but rather in spite of state intervention. Boulnois (1963) gave an account of Huang Ch'ao, who massacred 120,000 traders in China – Christians, Jews, Mazdeans and Muslims. The same author cited the Arab writer Abu Zaid Al-Hasan who wrote about China banishing Western merchants – Christians, Jews and Nestorians. In 1424, Emperor Hung-His of the Ming dynasty banned foreign expeditions, and in 1426 China closed its borders to keep out foreign influence. Emperor K'ang-hsi of the Qing dynasty (1644 to 1911) also banned foreign travel. Likewise, Chinese entrepreneurship developed without institutional support.

Rather than foster entrepreneurship, several emperors attempted to discourage the activities of entrepreneurs. Traditional Chinese law, influenced by Confucian principles, forbade merchants from wearing nice clothes in public. The law also forbade them from riding horses or from travelling on wagons.

The first Westerner to sail into China's Pearl Delta was Captain Jorge Alvares, in 1513. There, he found Guangzhou (Canton), already a great trading city at the time. In 1553, officials in Guang Dong accepted bribes from the Portuguese who wished to conduct trade in Macao. When the Portuguese discovered that the Japanese were willing to buy Chinese silk while the Celestial Empire prohibited business dealings between Chinese and Japanese merchants, Portuguese entrepreneurs prospered as middlemen between the two.

The Great Wall

The Summer Palace

Dynasty/Era	Period	Economy
Xia (Hsia)	2205 BC to 1766 BC	Slavery
Shang	1766 BC to 1122 BC	Slavery
Western Zhou	1122 BC to 770 BC	Slavery
Spring & Autumn Period	770 BC to 476 BC	Transition
Warring States Period	476 BC to 221 BC	Transition
Qin (Ch'in)	221 BC to 206 BC	Feudalism
Han	206 BC to 220 AD	Feudalism
Wei	220 to 265	Feudalism
Jin	265 to 420	Feudalism
Southern & Northern	420 to 589	Feudalism
Sui	590 to 617	Feudalism
Tang	618 to 907	Feudalism
Five Dynasties	907 to 960	Feudalism
Song	960 to 1279	Feudalism
Yuan (Mongol)	1279 to 1368	Feudalism
Ming	1368 to 1644	Feudalism
Qing (Manchu)	1644 to 1911	Feudalism
Modern Era	1911 to 1949	Bourgeois
Post-war Era	1949 to 1952	Rehabilitating
Planned Economy	1953 to 1978	Centrally planned
Opening Policy	1979 to 1992	Planned-commodity
Transitional	1993 to date	Socialist-market

Table 3.1. Historical Periods in China

In 1557, China allowed Portuguese merchants to establish homes and warehouses in Macao. The Portuguese supplied ivory from Africa and cotton from Goa, as well as cannons, clocks and mirrors from Europe. In exchange, Chinese entrepreneurs brought porcelain, seed pearls, and silk. Exporting silk from Macao to Japan proved to be highly profitable for the Portuguese entrepreneurs who were happy to be paid in silver. They then used the silver to pay for Chinese goods, which they sold in Europe.

During the 1600s, entrepreneurs from England decided to get silk and tea directly from China. A problem, however, was that Chinese consumers wanted nothing from England. This obstacle was resolved when Chinese entrepreneurs gave English traders silver in exchange for opium from India. The silver was then used to acquire silk and tea in China. Toward the end of the 18th century, the English also purchased from Chinese merchants a fine cotton fabric known as nankeen. In 1821, the English began using Hong Kong as a base for opium vessels.

The year 1839 marked the start of the Opium War between China and England. In 1841, the latter occupied Hong Kong Island, and in 1842 the Treaty of Nanking ceded Hong Kong to England. In 1860, Queen Victoria also acquired the Kowloon Peninsula and in 1898 leased the New Territories for a period of 99 years. In 1887, Portugal forced China to sign the Draft Agreement of the Sino-Portuguese Meeting. This was followed by the Sino-Portuguese Treaty of Peking, allowing Portugal perpetual administration of Macao.

In 1911, overseas Chinese entrepreneurs financed Dr. Sun Yat-Sen's bourgeois-democratic revolution, which overthrew the Qing dynasty. On December 25, 1914, Yunnan Province announced its separation from the Chinese Empire (Passantino, 1946). The Communist Party of China was established in 1921.

At the time, China was exporting up to 4,800 tons of silk annually. Things changed, however, when three quarters of its silkworms were found to be diseased in 1923 (Boulnois, 1963). Chinese sericulture was no longer a major force in international markets. In order to re-establish a significant market presence, the International Committee for the Improvement of Sericulture in China then purchased healthy silkworms in France and Italy. Yet, the 1920s were bleak. From 1927 to 1937, the Agrarian Revolutionary War took place.

The Sino-Japanese War – also known as the War of Resistance against Japan – started in 1937. Between January and September 1938, nearly 200,000 Chinese worked on the Burma Road, an all-weather highway linking Kunming, with Rangoon – as Yangon was called at the time. Outram & Fane described how the "opening of the Burma Road turned sleepy Kunming into a bustling metropolis. Capital and largest city of Yunnan Province, the ancient community slumbered half-forgotten until 1939. Even the opening of the narrow-gauge railway from Haiphong, French Indochina seaport, nearly thirty years before, failed to rouse it from lethargy. Suddenly trucks by the hundred poured into the city, which became a distributing center for the Chinese armies (1940, p. 654)." The Japanese occupation of Burma was tragic for China, as Burma had been the principal base for Allied aid to China (Christian, 1943).

During WWII, 30,000 Jewish Refugees Lived in Shanghai

Street Market in the Former Hong Kew Ghetto

The Peace Hotel, Formerly the Sassoon Building, in Shanghai

The Hong Kong & Shanghai Banking Corporation Building

Brown (1944), Stewart (1944) and Lowdermilk (1945) gave detailed accounts of enterprise in wartime China. When Hong Kong fell to the Japanese, many entrepreneurs fled from occupied Hong Kong to China. By the time the war was over, in 1945, millions of Chinese had been killed.

In 1945, capitalists in western China declared the independence of the Turkestan Republic. Following the National Liberation War – which lasted from 1945 to 1949 – on October 1, 1949, Mao Zedong proclaimed the establishment of the PRC. Turkestan was subsequently absorbed into this new entity.

From 1949 to 1952, the PRC focused on "the task of rehabilitating the national economy (Hong, 1990, p. 11)." The First Five-Year Plan, covering the years 1953 to 1957, gave rise to a centrally planned economy in China. Entrepreneurship in the PRC was eliminated, in favour of a system of state production and co-operatives. By mid-1950, about 20,000 refugees poured in every week from the PRC to Hong Kong (Long, 1954). Until 1978, business was production-oriented, based on a plan. Prices were fixed and advertising was banned by law.

The Chinese Model of Gradual Transition

In 1978, Deng Xiaoping declared that the PRC needed to "reform and open up," thus setting the nation on the road to economic liberalisation – with an open-door policy. The state moved away from Maoism, reduced its control of the economy, and business became increasingly oriented toward sales. Private enterprises – including foreign participation – were permitted. Some prices were allowed to float. Advertising was legalised.

In 1979, the State Council of China officially endorsed a policy allowing entrepreneurship to contribute to economic development, not as a replacement of central planning, but rather as a supplement to the socialist economy. Regulatory reform legalised entrepreneurship, and by 1980 there were over 1 million entrepreneurs in China.

The government also recognised *zhuanyehu* – "specialised households." These were families who were given permission to operate family businesses. Enterprises included animal husbandry, carpentry, construction, embroidery, and fish farming. Self-employed farmers were permitted to cultivate apples, beans, corn, grapes, pears, persimmons, rice, sorghum, soya beans, sugar beets and tangerines. A new constitution was adopted on December 4, 1982.

The Chinese proverb, "*A single spark can start a prairie fire,*" led to the "Spark" Programme, an important scheme that was launched in 1986, to promote entrepreneurship in rural areas of China. Through this programme, the State Science and Technology Commission encouraged the establishment of several thousand new ventures. Special incentives were offered to entrepreneurs who harnessed a technology deemed to be appropriate for rural industry. This included enterprises involving agriculture, aquaculture, food processing, light industry, textile manufacturing, and the production of components.

In 1992, the Fourteenth National Congress of the Communist Party of China proposed the establishment of a socialist market economy. During the early 1990s, annual growth reached 13.4%, as the orientation of firms was shifting toward the market. In 1993, the China Council for the Promotion of International Trade hosted a world conference in Beijing to encourage entrepreneurship. Devaluation of the yuan, in 1994, helped Chinese entrepreneurs increase their competitiveness in export markets. The state also introduced a variety of incentives for entrepreneurs who export. This included tax exemptions and easy access to financing via specialised banks.

In 1995, legislation declared 339 cities and counties open areas in which entrepreneurship could thrive, as a supplement to socialism. The Financial Security Law came into effect in October 1995. This legislation affected entrepreneurship inasmuch as it covered different forms of security: deposit, guarantee, lien, pledge and mortgage. Land-use rights and social facilities were placed on the list of assets, which could be mortgaged.

The All-China Federation of Industry and Commerce – with more than 80,000 members – established its Information Centre in 1995. Its major functions are: to provide assistance to foreign entrepreneurs, in finding Chinese partners for joint ventures; to perform market analysis; to provide consulting services to foreign entrepreneurs; to organise trade shows; to organise technology exchanges; and to assist foreign entrepreneurs. The federation also publishes periodicals of interest to entrepreneurs.

Until April 1996, foreign entrepreneurs benefited from Value Added Tax (VAT) exemptions and exemptions from customs duties on imported capital equipment. These were then withdrawn, thus putting foreign and local entrepreneurs on equal footing with regards to investment incentives.

Fast Food in Beijing

Weighing Bananas in Canton

In 1997, domestic growth in the PRC slowed down, and in 1998, the state responded by launching an elaborate programme of government spending on infrastructure. In 1998, the State Economic and Trade Commission of China was restructured such as to include a Department of Small and Medium Enterprises.

From a high of 43% in 1997, average tariffs were reduced to around 17%. Yet, imports remained subject to VAT. In contrast, local entrepreneurs were often exempted from sales tax. Thus, it may be said that the effective tariff rate is close to 40%, thereby protecting local entrepreneurs.

In July 2001, President Jiang Zemin made the news with his declaration that the Communist Party of China should recruit capitalists. He said this would boost the "influence and cohesiveness" of the party. That same month, China passed new rules on intellectual property.

A three-tier system exists in China today:

(1) Consumer staples, raw materials and industrial materials are supplied by the state and prices dictated by the government. This is truly a state-controlled planned sector.

(2) For other products – as determined by the state – prices can vary within a range dictated by *guojia wujia guanliju,* the State Price Bureau. This may be described as a hybrid sector.

(3) In contrast to the state-controlled planned sector, a free-market system allows the market to establish prices of some products.

LIU RONG YAO Manager

THE BUSINESS EXPLOITING GENERAL CO .OF
HENAN COAL GEOLOGY BUREAU
THE COAL TRANSPORTING AND SELLING GENERAL
STATION OF HENAN COAL GEOLOGY BUREAU

Add:No.202 Songshan Road
Zhengzhou China
Tel:(0371)448731 448734
Cable:0014
Postcode:450052

Bank of China Building, in Beijing

Driving to Work

Small Business in Beijing

Foreign Direct Investment

One Country, Two Systems

In 1984, the Sino-British Agreement announced the reversion of British Hong Kong to China, effective July 1, 1997. This was conditional on the PRC creating a Special Administrative Region, conforming to the concept of "One country, two systems." Hong Kong reverted to Chinese rule, but the Special Administrative Region of Hong Kong was allowed to keep its own legal and judicial system, as well as capitalist economy, until 2047.

Hong Kong Retained Its Currency

Along the same lines, in April 1987, Portugal signed the Joint Sino-Portuguese Declaration and Basic Law. This allowed the PRC to take over Macao, on December 20, 1999.

There are, nevertheless, important differences between Hong Kong and Macao. Although both were built on entrepreneurship, their industrial sectors evolved differently. Entrepreneurs in Hong Kong benefit from the government's laissez-faire policy, including favourable tax treatment. In contrast, beginning in 1981, Macao adopted an increasingly interventionist approach.

Hong Kong has at least 300,000 small and medium enterprises, which are flexible, and cost-effective. While their product mix is increasing, the typical size of an order is shrinking.

Macao Under Portuguese Administration

Hong Kong Star Ferry Terminal at Kowloon

Hong Kong Has the World's Largest Unsubsidised Public Transit System

The Western Market, in Hong Kong

Many entrepreneurs in Hong Kong are concentrated in the service industries. Others have manufacturing plants, outside Hong Kong. Entrepreneurs from Hong Kong have financed over 170,000 joint ventures in China, and these employ some 10 million people – twice the labour force in Hong Kong. According to unpublished sources at the Hong Kong Trade Development Council, two fifths of the exporters in Hong Kong, have operations in two or more economies.

The Trade Development Council of Hong Kong operates the SME Service Centre. This is a one-stop shop for entrepreneurs in search of information on technology acquisition and internationalisation.

In contrast, entrepreneurs in Macao tend to be small-scale producers of low-technology goods, including artificial flowers, ceramics, clothing, electrical products, electronics, firewood, footwear, furniture, machinery, optical devices, plastic goods, textiles and toys. Dana (1999e) provides a detailed contrast of differences between entrepreneurship policy and practices in Hong Kong and those in Macao.

Pluralism

While the Han-Chinese comprise the majority of China's population, the country is also rich with minority groups. There are 55 recognised national minorities in the PRC.

In Xinjiang, the largest administrative region in China, the Han-Chinese are a minority, and most people are Muslim. The indigenous people in Xinjiang are Uygurs – Muslims with Indo-European features.

A Muslim Uygur, in Xinjiang, sells *nan*, flat bread with *plov* – rice mixed with religiously prepared halal mutton – along with *shashlik* – halal mutton broiled on charcoal. On another street, Han entrepreneurs sell tofu, noodles, dumplings, dogmeat, fried vegetables, white rice and eggs cooked in tea. Pluralism is structural in nature, as members of the different ethnic groups do not share a secular mainstream arena. Whereas the Han-Chinese introduced a firm-type economy to Xinjiang, the non-Han sphere of activity is concentrated in the bazaar economy.

Muslim and Han in Xinjiang

Among the non-Han peoples in Xinjiang are Kazakh and Kirghiz minorities; these are also Islamic, with a Turkic language. In contrast to the Uygurs who have been cultivating the land, the Kazakhs and Kirghiz have retained their traditional nomadic lifestyle, which includes breeding camels and horses, both of these for meat and for milk as well as for transport.

Among the Kazakhs, horseback courtship is still practiced, as a suitor must kiss his woman, on the gallop. Although Kazakhs in Xinjiang practise agriculture on a limited scale, many are herdsmen, and their entrepreneurial spirit is strong. Clark (1954) described the thousands of families of Kazakhs in Xinjiang who opted to trek 3,000 miles, in order to escape communist control in China.

While the Hans speak Mandarin, the Uygurs speak a Turkic tongue, written with an Arabic script. The Uygurs have more in common with their Turkic neighbours of the formerly Soviet, Central Asian republics, than with the Han-Chinese. Perhaps because they are often looked down upon by Han-Chinese, the Uygurs do not generally identify with Chinese nationality, but rather they consider themselves to be under Chinese occupation.

Members of different ethnic communities, within Xinjiang, have distinctly different tastes. A particular favourite among the Hui is soup made of flour. In contrast to the Hans who use chopsticks, the Kazakhs in Xinjiang eat with their hands; these people enjoy smoked meat and horse intestines as well as mutton with noodles eaten without utensils. The Kirghiz prefer eating rice mixed with milk, and they exhibit a preference for horse's milk. A popular meal among Uygurs consists of hand-pulled noodles mixed with bits of lamb and vegetables, savoured with strong tea. They enjoy fried *sanzi* – made from dough – and a variety of steamed foods including buns and dumplings; they also like roasted cubes of mutton on skewers. They eat *zhuafan* with their fingers and their favourite cake is *nang*. For the Xibe people, a special meal would include sheep entrails. The Mongolians, the Uygurs and the Xibe drink milk-tea. As for the Kazakhs and the Kirghiz, their favourite drink is fermented mare's milk, known in Central Asia as *koumis*, and in China as *manaizi*. In major centres, Western food has also made inroads, with chains such as Kentucky Fried Chicken (Dana, 1999d). Members of each ethnic group, in Xinjiang, appear to stick to themselves, and each is physically distinguishable. Rather than wearing Chinese slacks, Uygur women sport colourful dresses or skirts with matching, brightly coloured scarves. Among the Uygurs, both men and women wear a colourful head cover. Daggers are fashionable among Uygur, Kazakh, Kirghiz and Mongolian men.

Arabic Script Above Chinese

Fresh Noodles Drying in the Sun

Eating Halal

Devout Muslim in Xinjiang

The Uygurs continue to cultivate cotton, grapes and melons. Local green grapes are still dried in the traditional manner, producing export-quality raisins with no pits. At local markets, such as that in Turpan, donkey- carts wait for local loads. Closer to the Kyrgyz Republic, trucks at the Kashgar market represent major international business. While some Uygurs have opted to urbanise, many have decided to remain in rural areas, residing in rectangular farmhouses with scuttles on flat roofs.

Recently, a new class of international merchants has been emerging with a pan-Turkic scope. Rather than doing business with Han-Chinese, the Muslims of Xinjiang express a preference to trade with co-religionists from the Turkic republics of the former Soviet Union. While the majority of the world's international trade takes place in the firm-type economy, Uygur traders thrive in the bazaar.

Perhaps the most striking bazaar is at the Horgas Pass, to the south-west of Bortala. Like other towns of the Silk Road[6], the settlement was impoverished when caravans stopped passing through it. In 1971, all international activity was banned here, as the border with the Soviet Union was closed. The local inhabitants had no electricity, no running water, and no trade.

Spices of the Silk Road

[6] This term refers not to one road per se, but to a web of inter-weaving paths that linked towns and bazaars.

In 1983, the border was re-opened. Entrepreneurs carried 20 million kilograms of goods across this border in 1984, according to information obtained by the author during an interview in Kazakhstan. Trade was predicted to surpass 500 million kilograms in 2003.

The demise of the USSR meant little to the average Chinese individual. However, the independence of the five formerly Soviet Central Asian states – Kazakhstan and the Kyrgyz Republic in particular – had a great symbolic lift for Uygurs, for Kazakhs and for Kirghiz living in Xinjiang. These Turkic-speaking minorities had historical, religious and cultural links with their Turkic-speaking neighbours in the Commonwealth of Independent States.

Uygur entrepreneurs in Xinjiang began to rebuild commercial links between China and Central Asia. In Khorghus, formerly a Silk Road town, now on the Kazakh side of the Kazakhstan-Xinjiang border, a new bazaar is sprawling. A unique feature along the border of China and Kazakhstan is "The Bazaar for Entrepreneurs from Both Sides." This market has an entrance in Xinjiang and another in Kazakhstan. Uygur entrepreneurs from Xinjiang bring beer, vodka, sugar, candies, clothing, leather goods, sporting apparel, toys, house-wares, tools, cigarettes and jewellery, all of which they sell to buyers from Kazakhstan, including Kazakhs and Russians, most of whom arrive by bus. Thus, pan-Turkic trade is on the rise, as Muslims in Xinjiang conduct business with Muslims in republics of the CIS. Although the bazaar is not simply a means to monetary reward, it is producing wealth among the Uygurs. More importantly, it allows Uygurs to socialise among others sharing similar cultural values. The flow of commerce is still fragmented into numerous transactions between individuals, just as it was generations ago. This allows the sharing of cigarettes over tea, and it helps to spread risks.

Officially, all of China has only one time zone, which is that of Beijing. Nevertheless, Uygurs function according to their own unofficial time zone. Uygur entrepreneurs, in Xinjiang, are more concerned with conducting business with Kazakhstan and the Kyrgyz Republic, than with pleasing politicians in China's capital; although Han-Chinese in Xinjiang respect Beijing time, most Uygurs set their watches two hours behind the official time of the PRC. This presumably facilitates pan-Turkic trade between Uygurs and entrepreneurs in Kazakhstan and in the Kyrgyz Republic; more importantly, it is a highly symbolic gesture.

Given that the Uygurs have a Turkic culture and that they speak a Turkic language, this has facilitated trade with bazaar entrepreneurs in the rapidly developing, newly independent republics of Central Asia.

Street Scene in Turpan

Prices at the Market

Toward the Future

Across the PRC, entrepreneurship has become a supplement to socialism and industrialisation is viewed as a complement to agriculture. Unlike other economies, which rapidly abandoned communism, the PRC is liberalising its economy slowly, thus avoiding spiral inflation. As well, its township enterprises are helping the nation industrialise, while avoiding uncontrolled urbanisation. In the longer term, sustained growth will likely require privatisation of state enterprises and a free flow of labour.

Meanwhile, local and foreign entrepreneurs are playing an important role in the economic and social development of China, despite tight control by Beijing. Heavy investment has been pouring in from overseas Chinese entrepreneurs, and the PRC today has the world's second largest GDP. With membership in the World Trade Organisation, China's model of transition is likely to bring increasing prosperity to its people.

KFC in Shanghai

The China World Tower, in Beijing

Chapter 4

West of China: An Overview of Central Asia

Introduction

With the demise of the USSR in 1991, five new countries became independent in Central Asia – the region between China and Russia. These are Kazakhstan, the Kyrgyz Republic, Tajikistan, Turkmenistan, and Uzbekistan. The area holds an estimated 200 billion barrels of untapped oil.

Whereas independence from the Soviet Union enabled the economies of the Baltic States – that had a history of capitalism and independence prior to communist rule – to flourish with a prosperous firm-type sector, the situation in Central Asia reflects a variety of different models. O'Driscoll, Holmes & Kirkpatrick (2001) described Turkmenistan and Uzbekistan as "repressed" relative to Kazakhstan, the Kyrgyz Republic and Tajikistan.

While the indigenous peoples of this region share Islam as their predominant religion, important differences must not be ignored. Kazakhstan uses the Roman alphabet; in some provinces of the Kyrgyz Republic, Roman script is prevalent, while Cyrillic is used in others. In Tajikistan, Arabic script is used. While independence allowed Kazakhstan and the Kyrgyz Republic to pursue unprecedented economic reforms, many Uzbeks swayed more toward religion. With the exception of Tajikistan, where independence resulted in civil war, the Islamic Renaissance Party was outlawed throughout the region. During the war, the Islamic Movement of Uzbekistan (IMU), a guerrilla organisation, fought with the United Tajik Opposition. Before retreating to Afghanistan, the IMU declared its intention to establish a caliphate – a religious government – in the Fergana Valley. In Afghanistan, the IMU assisted the Taliban's campaigns against the Northern Alliance. Kazakhstan and Turkmenistan, meanwhile, escaped the attention of Taliban-backed groups.

The Kyrgyz Republic proved itself right at the forefront of economic reform. Germany, Japan, the Netherlands, Switzerland, Turkey, the United States, the European Bank for Reconstruction and Development, the European Union, the International Monetary Fund (IMF) and the World Bank poured assistance into this new country. The United States Agency for International Development (USAID) sent refined vegetable oil, while German buses arrived to replace older Soviet-built models. Foreign firms scrambled to invest in the rapidly developing republic. The paradox, however, is that the process resulted in economic polarisation. Although the oil received from USAID is labelled "NOT TO BE SOLD OR EXCHANGED," it is sold at prices beyond the means of many, as are the new bus fares.

Officially Not Available for Sale, But Sold Anyway

While it is important to distinguish *among* the various countries of this vast region, it is also useful to recognise the cultural differences that reflect the pluralism *within* each of these republics. Ethnic solidarity, in Central Asia, keeps the indigenous Muslims segregated from the Russian minority. This is similar to the situation in Xinjiang, where members of the Muslim community have minimal interaction with the Han-Chinese, as seen in Chapter 3.

Bonacich & Modell (1980) examined the economic basis of ethnic solidarity among Asians in the melting pot scenario of the United States. This contrasts sharply with the situation in Central Asia, where structural pluralism is reinforced by the fact that different communities have distinct languages – with different alphabets – and incompatible dietary restrictions – arising from vastly different religions. The Kazakhs and Uygurs in China do *not* share a secular mainstream arena with the Han-Chinese; there is minimal interaction across cultures. Likewise, the Russian minority in the Kyrgyz Republic does not mix with the Uzbek minority in that country. Rather, each ethnic group has its distinct institutions and attitudes toward business. Herbig & McCarty suggested, "unlike the innate positives inherent in the Chinese culture, the Russian (Slavic) culture does not have underlying innovative tendencies or an entrepreneurial history (1995, p. 48)." The same authors elaborated:

> Historically, the Russians have been anti-capitalist. During Czarist Russia, the land owning nobles and state officials who governed the empire for the most part looked down on capitalists and decried the profit motive. So too did the intellectuals. Capitalism was denounced as materialist, selfish, and above all Western in spirit (Herbig & McCarty, 1995, p. 53.)

In contrast, the indigenous peoples of Central Asia had a long history of self-employment and commerce along the Silk Road – and this was so until central planning was introduced by the Soviets. Likewise, the Muslim Uygurs of western China thrived in their bazaar economy – until central planning was imposed from Beijing. Not surprisingly, the values and entrepreneurial behaviour expressed by these people contrast greatly with their Russian neighbours.

Historical Overview

Recent findings reveal that a major society thrived in Central Asia, 4,000 years ago, with a sophistication that perhaps exceeded those of the Indus Valley or Mesopotamia. Its industrious dwellers constructed buildings that are larger than the biggest structures of ancient Mesopotamia or China. Ruins, including elaborate works in bronze and marble, indicate that raw materials were imported.

In 1904, Raphael Pumpelly, an archaeologist from New Hampshire, discovered the ancient ruins at Anau, but there was little follow up for a long time. Exploring the region during the 1970s, Soviet researchers discovered the existence of several hundred structures, each one larger than the size of a football field, and surrounded by a series of walls. Although these findings were published in Russian journals, they were not translated. It was during the 1980s that this literature was first read by Professor Carl Lambert-Karlovsky, of Harvard. In 1988, Dr. Fred Hiebert – a Harvard graduate and former student of Lambert-Karlovsky – went to Central Asia, to study ancient trading practices along the Silk Road. In desert oases currently situated in Turkmenistan and Uzbekistan, he discovered evidence of a culture with a unique written language. Settlements along the oases served as nodes along a trading route that linked China with Mesopotamia. Local people irrigated desert land in order to grow wheat and barley. They carved imported marble into fine art and they forged metal.

Approximately 1,500 years ago, the Silk Road came to be used by camel caravans travelling between China and the Middle East. It was the establishment of sophisticated maritime routes that resulted in the gradual decline of trade along the Silk Road.

Central Asia came under Russian rule during the 18th century, when the tsars sent Cossacks – descendants of Russian serfs who escaped from their estates – to establish outputs for the purpose of protecting trade routes between Europe and the Far East. Tsarist repression of Central Asia during imperial times prompted popular support for the social democratic revolution that hinted at self-determination for the region. However, Jewish and Muslim locals were deliberately excluded from the regional government, which was set up in Tashkent,[7] known in former times as Chach. This allowed the Bolsheviks to seize control of Central Asia, in November 1917; a few months later, they renamed themselves communists.

---

[7] Tashkent literally means "castle of stone."

Between 1918 and 1920, the communists attempted to abolish private commercial activity. Nonetheless, barter and wages-in-kind were specifically permitted, and while officially outlawed, black market trade was tolerated (Prasso, 2001).

In 1921, Vladimir Ilyich Lenin introduced his New Economic Policy, which exempted some private enterprise – including traditional farming – from central planning. Peasants were permitted to own land and small-scale entrepreneurs were tolerated in industry and in trade.

In April 1921, Soviet Turkestan became an autonomous Soviet socialist republic, within the Russian Federative Socialist Republic. At the time, the nationalities of Central Asia did not exist as such; people were known as either Persians or Turks. Soon, decision-makers in Moscow would create new republics. In 1924, a Soviet border commission reorganised Central Asia on a national basis; new republics were invented by the stroke of a pen. Uzbekistan was thus created in 1924. On February 1, 1926, the Kirghiz homeland – carved out of Turkestan – became an autonomous Soviet socialist republic, within the Russian Federative Socialist Republic. The man-made borders did not correspond to ethnic or linguistic lines; although people in Bukhara – not to be confused with the Khanate of Bokhara – and Samarkand spoke Tajik, these cities ended up in the Uzbek Soviet Socialist Republic.

In 1928, Josef Stalin replaced Lenin's relatively tolerant policy and imposed a Five-Year Plan that imposed collectivism. Yet, resistance was strong in Central Asia, notably among Kazakhs and Kirghiz, whose traditional nomadic lifestyle was altered. While economically deprived peasants in rural Russia were happy to rid themselves of feudalism, this was not an issue in Central Asia where trade along the ancient Silk Road had brought prosperity to capitalist territory.

In 1929, policy-makers in Moscow created the Tajik Soviet Socialist Republic. The Turkmen Soviet Socialist Republic was formed in 1934, followed by the Kazakh Soviet Socialist Republic and the Kirghiz Soviet Socialist Republic in 1936.

A result of Stalin's anti-religion campaign – which began in 1932 – by 1940, 25,000 mosques had been destroyed and 14,500 Islamic schools were closed down. When the Nazis invaded the Soviet Union, in 1941, Soviet Germans – citizens of the USSR – were deported from Europe and relocated to Central Asia. In an act the Soviets called "social engineering," the demographics of Central Asia were changed, and the effects are still visible.

In 1949, the victory of the communists, in China's civil war, boosted belief in the communist doctrine. During the 1960s, however, animosity developed between China and the Soviet Union; in 1969, their forces fought along their common border. Families of Central Asians were caught on opposite sides of the international boundary.

While China tried to bring the Muslims of Xinjiang ideologically closer to Beijing, Moscow continued its attempt to Sovietise Central Asia. Whereas the Arabic alphabet had been that used in the region, the Soviets legislated Cyrillic to replace it. The five Soviet republics of Central Asia depended on Moscow for planning and for direction.

During many years under central planning, the story was told of a Communist Party chief who was said to have asked a farmer, "How much cotton do we have?"

"Enough to pile up to Allah's feet," was the farmer's response.

"We have no God here!" exclaimed the communist.

"OK," responded the farmer, "We have no cotton either."

In 1991, the breakdown of the Soviet bloc disrupted traditional trade links and distribution channels. Since then, pan-Turkic sentiment resulted in the development of trade and investment links between Turkey and Central Asia. Following the independence of the Central Asian republics from Moscow, there was a mass emigration of Germans, Jews, and Slavs, from the region.

On Dec 21, 1991, at a meeting in Almaty – the capital of Kazakhstan at the time – all five ex-Soviet countries of Central Asia joined the CIS, which in 1993 came to include all of the formerly Soviet republics with the exception of the three Baltic States, namely Estonia, Latvia and Lithuania.

In January 1993, all except Tajikistan agreed on a charter to establish an economic co-ordination committee. On September 24, 1993, all except Turkmenistan signed an agreement to form an economic union with Russia and other CIS members. In January 1994, Kazakhstan, the Kyrgyz Republic and Uzbekistan agreed to create a single economic zone. In October 1994, all the members of the CIS except Turkmenistan established the Inter-Government Economic Committee, to co-ordinate energy, transportation and communication policies. On March 29, 1996, Kazakhstan and the Kyrgyz Republic joined Belarus and Russia in the Community of Integrated States.

Pluralism & the Next Five Chapters

A comprehension of historical factors, pluralism and cultural differences might enhance our understanding of the future of transition, especially in this part of the world. Most importantly, the region must not be mistaken for homogeneous. Traditionally, the Kazakhs were pastoral nomads, while the Uzbeks engaged in sedentary agriculture. Under Soviet rule, the Kazakhs were more Russified than was the case with other nations; today, the Kazakhs adhere relatively loosely to Islam. In the Kyrgyz Republic, the people of the south are generally more observant of Islam, and less Russified, than are the northern Kyrgyz. While the Tajiks – Indo-European in origin, with an Iranian tongue – might be said to emulate Iran, the Kazakhs – of Altaic stock with a Turkic language – appear more interested in economic transition, than in religion.

The next five chapters will look at each of the Central Asian republics of the CIS, namely Kazakhstan (Chapter 5), the Kyrgyz Republic (Chapter 6), Tajikistan (Chapter 7), Turkmenistan (Chapter 8), and Uzbekistan (Chapter 9). Research for these chapters was conducted by the author in Central Asia.

Faces of Central Asia

Chapter 5

The Republic of Kazakhstan[8]

Introduction

On December 16, 1991, Kazakhstan, formerly known as the Kazakh Soviet Socialist Republic, became the last of the Soviet republics to declare its independence. The new country became the ninth largest in the world, covering 2,717,300 square kilometres; this area is comparable to that of Western Europe – approximately half that of the continental United States. Kazakhstan borders China, the Kyrgyz Republic, Russia, Turkmenistan, Uzbekistan, the Aral Sea, and the Caspian Sea. The republic is the third largest industrial power in the CIS. In terms of natural resources, per capita, Kazakhstan is perhaps the richest in the world. Kazakhstan has the world's third largest oil reserves, after the Persian Gulf and Siberia.

During the early 1990s, there were doubts as to whether the republic would embrace a Western-style market system. In September 1991, *The Economist* suggested that conditions could lead to Islamic fundamentalism. Instead, Kazakhstan became an example of successful transition to a relatively open, capitalist economy, consistent with the traditional cultural values of the Kazakh people. O'Driscoll, Holmes & Kirkpatrick (2001) found less government intervention in this country, than anywhere else in Central Asia. The republic emerged as a vanguard of reform and privatisation. Annual inflation, which reached 176% in 1995, was controlled to 7% in 1998, before rising to 13.8% in 2000. In January 2002, the government resigned saying it was time to make way for people with new ideas.

---

[8] This chapter includes information obtained from: the Ministry of Agriculture; the Ministry of Economics; the Ministry of Finance; the Ministry of Foreign Affairs; the Ministry of Industry and Trade; and the Ministry of Internal Affairs. The chapter includes material that first appeared in a refereed journal article about Kazakhstan (Dana, 1997a).

Historical Overview

The Kazakhs – a Turkic people, with Mongol features – claim to be descendents of the hordes of Genghis Khan. These people have historically been self-employed nomads. They roamed a vast terrain of desert and mountains, and they lived off their capital, breeding and raising livestock for a relatively comfortable subsistence, while accumulating considerable wealth in the process. Their horses were regarded not only as priceless servants, but also as a status of wealth.

The Kazakhs fell under Russian rule during the 18th century, when the Cossacks arrived to protect trade routes. During the 19th century, peasants arrived from Russia and the Ukraine. In 1882, Muslim Dungans settled here, escaping religious persecution in China.

Following the Bolshevik Revolution, many Kazakhs fled to Mongolia (Edwards & Ludwig, 1993). Greed on the part of the Red Army – which requisitioned food – reinforced hostility against communism and resulted in support for a new alternative, pan-Turkic nationalism. Lenin's response was to hire Enver Pasha – son-in-law of the caliph – to pacify the area. An over-achiever, Enver Pasha attempted instead to create his own empire.

Hoping to obtain some political power, the indigenous people would try their luck with the Communist Party of the USSR. Yet, they did not really adopt communism as an ideology, as it was not compatible with their cultural belief system.

Although some changes took place relatively peacefully, the transition to collectivism did not. The collectivisation of land – which began in 1928 – evolved into a trauma as Kazakh herdsmen slaughtered their livestock rather than surrender them to collective farms. In 1933, 32 million horses, cattle, goats and sheep were wasted.

On December 5, 1936, the Kazakh Soviet Socialist Republic was carved out of Russia. This was the last of the Central Asian republics created by Moscow. Only 25% of the Kazakhs survived the new regime. The Soviets introduced social engineering, repopulating the land with prisoners from across the USSR.

In 1941, Russia and the Ukraine were home to thousands of Germans, descendants of people who had been invited – by Catherine the Great – during the 18th century. When the Nazis invaded the Soviet Union that year, Soviet Germans were deported to Kazakhstan, where they eventually numbered 1 million people.

Moscow Introduced the Russian Language to Kazakhstan

Commemorating the Victory of the USSR in WWII

After World War II, the Politburo in Moscow attempted to transform the Kazakh Soviet Socialist Republic into a breadbasket. Wheat was planted, but the plan failed as the semi-arid steppe – covering 49% of the world's tungsten – could not support long-term farming.

The Soviets then decided to use the country as a missile testing range and nuclear test site. Between 1948 and independence in 1991, atomic bombs were exploded here, at an average of one every three weeks.

The Soviets also imposed cotton growing in Central Asia. For this, however, they needed water, and so they diverted the rivers that once flowed into the Aral Sea, in order to irrigate land for cotton (Edwards & Ludwig, 1993). Simultaneously, chemical fertilisers for the cotton made their way into the Aral.

The Aral Sea – once the fourth largest inland body of water in the world – was the largest body of water between the Caspian Sea and the Pacific Ocean. It was home to a couple of dozen species of fish, and 60,000 fishermen derived their livelihood from this. A problem, however, was that by diverting the rivers which fed the Aral Sea, the sea was deprived of replenishment, and since 1960, it lost almost 75% of its volume. In 1987, the sea split into two separate bodies of relatively shallow water. Not surprisingly, 60,000 fishermen lost their jobs. The former seabed became a salt desert from where winds pick up salt and sand; up to 150 million metric tonnes of salt and sand are then dumped on once-fertile land. The irony is that by diverting rivers in the attempt to improve soil productivity, the ecosystem was altered, such that winters became colder and last a month longer than in the past. Simultaneously, fertile farmland is being contaminated by the residue of a disappearing sea – that has receded 50 kilometres from the Kazakh port of Aralsk.

The history of mankind knows no other example where, before the eyes of a single generation of people, an entire sea disappears from the face of the earth

– P. Khabibulaev (about the Aral Sea)

Not From the Aral Sea

Cold Winters

The Kazakh Model of Transition by Decree

In 1990, Kazakh replaced Russian as the official language in the Kazakh Republic, and Kazakhstan became an independent nation on December 16, 1991. The name of the national capital, known as Verney under Russian rule, and later as Alma-Ata, was changed to Almaty – literally "Father of Apples." In January 1992, the government opened up foreign trade.

In January 1993, the state introduced the National Council for Economic Reform. A privatisation plan scheduled the sale of most state enterprises with more than 200 employees. A new Kazakh currency – the tenge – was introduced in November 1993, replacing the Russian rouble.

The New Currency

In January 1994, Kazakhstan signed an agreement to create a single economic zone with the Kyrgyz Republic and Uzbekistan. In April 1994, the privatisation of 3,500 medium firms took place by auction, and more reforms were instituted during the year. Privatisation was comprehensive and liberalisation proved extensive. Small enterprises with fewer than 500 employees, and large firms with over 2,000 workers were sold for cash.

In December 1994, the Law on Foreign Investments introduced incentives for foreigners, and unpublished sources of the Ministry of Economy indicate that by 1995, 2,000 joint ventures had been created. A comprehensive structural reform programme was implemented, and by the time a new customs code liberalised trade in July 1995, the monthly rate of inflation had fallen to 2%. Over 100 banks came to operate in Kazakhstan. These included Agroprom Bank; Alim Business Bank; Albaraka-Kazakhstan; Alembank; Alfa Bank; Alkom Bank; Almaty Trade-Finance Bank; Altyn-Dan Bank; Aralecobank; Asia Kredit; Asiaturbank; Bank of Texas and Kazakhstan; Citibank; HSBC; Ilan Bank, and others. Of course, not all survived.

The transportation infrastructure was also rapidly developed, with several airlines replacing the Aeroflot monopoly. These included Aeroservice, Kazakhstan Aue Zholy, Luftbruke and Sputnick. Unlike other formerly Soviet republics, Kazakhstan has a good supply of energy, and therefore, when other airlines were grounded due to fuel shortages, Kazakh Air managed to fare relatively well. Tourism is becoming developed. Some railway stations operate *komnaty otdykha*, inexpensive rooms to spend a night. Hotels and restaurants are sprouting up around the country, and service has been improving.

A priority for the state has been the ambitious privatisation programme. *Perestroika* is highly visible here. In contrast to the Soviets who attempted to industrialise Kazakhstan in frenzy, the post-independence government welcomed foreign capital, technology and expertise to co-operate in a co-ordinated development effort. Chevron signed an agreement to extract oil, and British Gas expressed interest as well.

Commercial stores identified as *Kommercheski Magazin* offer imports from the West, while Chinese merchandise is available at the markets. In Almaty, one finds hundreds of traders from China. In Chorgos – eastern Kazakhstan – a Chinese bazaar has been set up. This is not to suggest, however, that Kazakhstan is without problems.

Many people complain that the worst of Western values have been imported, including crime and racketeering. Most respondents complain of government corruption, greed and unemployment.

In December 1998, Kazakhstan shifted its national capital from Almaty to Akmola, which literally means "capital." The name of the new capital was later changed to Astana. The purpose of the move was to set a new era, and the move took place ahead of schedule, but Soviet bureaucracy survived. This is seen as providing jobs for otherwise useless individuals. Regulations exist, but rules are bent. The law stipulates that the tenge is the legal tender currency, but in practice, the US dollar is used with equal ease.

Devaluation of the Kazakh tenge, in 1999, increased the competitiveness of exports. A rise in unemployment, however, prompted an exodus of Germans and Russians from Kazakhstan (Sealy, 2000).

Strategy 2030 is the economic development strategy to make Kazakhstan "a Central Asian snow leopard," – a model distinct from that of the Asian tigers.

Russian Imports Are Popular

Entrepreneurship

Clark (1954) described entrepreneurial activities of Kazakhs at the time, but until *perestroika*, entrepreneurship was not officially allowed in the Kazakh Soviet Socialist Republic. With transition, entrepreneurship was legalised, and in 1995, the Tax Code of July 1, 1995 specifically decreased the tax burden of entrepreneurs. At the time, there were 32,186 small firms in Kazakhstan, providing 330,000 jobs.[9]

As noted by Zhuplev, Kiesner, Kozhakmetov, Tan & Konkov (1998), Kazakhstan has been far more successful than Russia at developing entrepreneurship. Given that small enterprises in Kazakhstan tended to be concentrated in trade and intermediary activities, new laws attempted to promote small-scale industry in agriculture and food processing; banking; construction; consumer goods; furniture production; glassware production; light manufacturing; medical tools and pharmaceuticals; paint and varnish production; primary wool processing; sewing items; and sheepskin tanning and colouring. Kazakhstan became a world leader in the production and export of pomegranate juice.

Micro-enterprise was also allowed to evolve with minimal interference. Unofficial vendors board passenger trains to sell beer, milk and snacks among a variety of refreshments. Along sidewalks, individuals set up tables and sell limited assortments of goods. Typical of the bazaar economy, prices are a function of time, cost, negotiating skills and the apparent wealth of a potential customer.

Pluralism

Following the independence of Kazakhstan, thousands of ethnic Kazakhs who had been living in Mongolia migrated to their new country. By 1993, the Kazakhs accounted for 40% of the population of Kazakhstan, outnumbering the Russians who comprised 38% (Edwards & Ludwig, 1993).

---

[9] These are unpublished data provided by the Congress of Entrepreneurs, 597 ul. Seifullin, Almaty.

Successful transition in Kazakhstan prompted thousands of ethnic Kazakh families residing in Iran, Russia, Turkmenistan and Uzbekistan, back to their motherland. In 1994, 280,000 Russians emigrated from Kazakhstan.

In 1996, Kazakh President Nursultan Nazarbayev announced that up to 4,000 Kazakh families would receive financial assistance to relocate to Kazakhstan. Ethnic Kazakhs soon accounted for more than 50% of the nation's population.

In 2001, the Republic of Kazakhstan was home to almost 17 million people, of whom 57% were urban dwellers. The annual population growth rate was 0.03%, the slowest in Central Asia, and extremely favourable when compared to that of Tajikistan (2.12%). Only 27% of the population of Kazakhstan was under 15 years of age, compared with 41% in Tajikistan.

Despite initial fears that Islamic fundamentalism could take over the nation, this seems highly unlikely in the near future. During Soviet rule, the Muslims in the Kazakh republic were reduced to a minority – of 41.9% – outnumbered by Russians and Ukrainians; although many Russians have departed for Russia, Kazakhstan is still a multicultural society with Germans, Jews, Koreans, and Crimean Tatars among others. The nation is relatively less Muslim than the other Central Asian republics, and religion is not a major concern here.

Toward the Future

Kazakhstan became the world's fourth nuclear power only because the Soviets felt the country's million square miles of steppe – 90% of the republic's surface area – were ideal for placing nuclear warheads. After decades of imposed communism, horses once again roam freely among Kazakh shepherds tending their flocks and drinking fermented mare's milk. Nearby, at the Baikonur Cosmodrome, satellites are being sent into orbit, $1.5 million per launch – a price considerably lower than the $70 million being charged by NASA. A 20-year lease, ending in 2015, allows Russia the use of 6,000 square kilometres of Kazakh land, including the city of Baikonur – formerly Leninsk.

While some of the other formerly Soviet republics are blaming their present on the past, Kazakhstan is focusing on the future. In contrast to the situation elsewhere, local cultural values in Kazakhstan are compatible with capitalism and with the re-emerging entrepreneurial spirit. As other republics may look at religious fundamentalism for leadership, and yet others yearn for a return to power of the Communist Party, Kazakhstan is keen on entrepreneurship, innovation and change.

The Author With Tatar Interpreter in Kazakhstan

Chapter 6

The Kyrgyz Republic[10]

Introduction

The Kyrgyz Republic is 198,500 square kilometres, situated in the Tien-Shan Mountains. It neighbours China, Kazakhstan, Tajikistan, and Uzbekistan. O'Driscoll, Holmes & Kirkpatrick (2001) found the Kyrgyz Republic to have less black market activity than any of its neighbours. The same report described the Kyrgyz Republic as more welcoming to foreign investment, than any other country in Central Asia. While joint ventures have boosted employment in the nation's capital, efforts to expedite transition have diverted attention from indigenous entrepreneurship.

Historical Overview

The area known now as Kyrgyzstan was historically a land upon which 40 Kirghiz tribes led nomadic lifestyles. It was annexed by Russia in 1864, and Russians and Ukrainians introduced sedentary agriculture to the area. Repression of Muslims in China prompted many Dungans to immigrate to Kirghiz lands in 1882. In turn, repression of Muslims, by Russia, during Czarist rule, prompted a Kirghiz rebellion in 1916.

[10] This chapter includes information obtained from: the Ministry of Agrarian Policy; the Ministry of Agriculture and Foodstuffs; the Ministry of Culture; and the Ministry of Industrial Policy. The chapter also includes material that first appeared in a refereed journal article about the Kyrgyz Republic (Dana, 2000a).

Traditional Kirghiz Home

Russian-style House in the Kyrgyz Republic

In November 1917, the Bolsheviks seized control of the area, and in 1918 the Kirghiz homeland became a part of Turkestan – which became an autonomous Soviet socialist republic, within the Russian Federative Socialist Republic. In 1924, a Soviet border commission began implementing a policy of divide and rule in Central Asia, and the Kara-Kyrgyz autonomous region was carved out of Turkestan.

During the ten-year period, from February 1926 to December 1936, Kirghizstan was an autonomous Soviet socialist republic within the Russian Federative Socialist Republic. In December 1936, the Kirghiz Soviet Socialist Republic was declared one of the constituent Soviet socialist republics of the USSR. It was commonly referred to as Soviet Kirghizia.

In 1928, the script of the Kirghiz language was changed from the Arabic alphabet to the Roman one. Collectivisation also began in 1928. This put an end to the traditional nomadic lifestyle of the Kirghiz, but resistance to collectivism was exceptionally strong; rather than transfer their flocks to collective farms, many Kirghiz herdsmen slaughtered their sheep – as did the Kazakhs. In 1941, the nation's script was changed yet again. Cyrillic was imposed on Soviet Kirghizia, replacing the Roman alphabet.

The Kyrgyz Model of Transition by Decree

In 1990, a law on ownership was introduced in Soviet Kirghizia, followed by a law on enterprises in 1991. In August that year, the Communist Party was outlawed. Independence was declared on August 31, 1991 and the Kirghiz people adopted a new name for their country – Kyrgyzstan – officially referred to as the Kyrgyz Republic.

The Economist, in September 1991, suggested that Central Asia could fall to Islamic fundamentalism; instead, the Islamic Renaissance Party was outlawed in the Kyrgyz Republic. Pragmatic people, the Kirghiz are much less preoccupied with religion than are the Uzbeks. While bus passengers in neighbouring countries listen to passages of the Koran, passengers on Japanese-made buses in Kyrgyzstan watch American movies on board.

In December 1991, a legal document (Kyrgyz Republic, 1991) declared that, with the exception of the production of currency, drugs and weapons, entrepreneurial activity is open to all people in Kyrgyzstan, regardless of nationality. In 1992, prices were liberalised and monthly inflation reached 30 to 50%. Also in 1992, the Roman alphabet was re-introduced.

The Russians Introduced Cyrillic

Soviet Industrial Centre on Lake Issyk-Kul

A new constitution was adopted on May 5, 1993. As part of an IMF anti-inflation scheme, a Kyrgyz currency was introduced on May 10, 1993. Known as the som, it allowed Kyrgyzstan to be the first of the Central Asian republics to abandon the rouble. By mid-1993, inflation had been reduced to a monthly average of 17%. The som established itself as one of the most stable, freely convertible currencies in the region. Unpublished sources, at the State Property Fund of the Kyrgyz Republic, claim that inflation fell from 1,360% in 1993 to 87% in 1994. Between 1994 and 1999, there were periods during which the som appreciated against the US dollar. However, during the late 1990s, about half of all families were living below the poverty line, defined by the government as approximately $32 per month.

The New Kyrgyz Currency

Meanwhile, a bold privatisation scheme that had begun in spring 1993 led to the sale by auction of enterprises with less than 100 employees. Larger ones were transformed into joint stock companies. In May 1993, the new Law on Foreign Investments introduced tax incentives. By 1994, there were 370 joint ventures in Kyrgyzstan, with Canadians providing the largest investment. In July 1994, eleven local entrepreneurs contributed $5,000 each, and established the Kyrgyz stock exchange. It was the first privately owned stock exchange in Central Asia.

Reform in the Kyrgyz Republic also encouraged the development of an elaborate banking infrastructure. Recent joint ventures include the German-Kyrgyz Orient Industrial Bank and the Kyrgyz-Swiss Joint Commercial Bank "Adil." Other banks include the Agroprombank, Bishkek Joint Stock City Bank, Commercial Bank Maksat, Hong Kong-International Commercial Bank Henfen, Kyrgyzautobank, and Mercury Bank Ltd.

Paying Before the Price Changes

Foreign investors were welcomed and joint ventures created jobs in the national capital, Bishkek – formerly known as Frunze. Among these was a Turkish firm, which participated in the introduction of Coca-Cola in 1996 (Dana, Dana & Vignali, 1999).

The Kyrgyz Republic adopted its Foreign Investment Law on September 4, 1997. In 1998, the government legislated the right to own land. On July 17, 1998, the country became the first of the region to join the World Trade Organisation (WTO). In 1999, guerrillas from the Islamic Movement of Uzbekistan attacked the republic. After the War Against Terrorism, a tent city was built for US soldiers near Bishkek.

Meanwhile, rural entrepreneurship was neglected. Although agricultural land was promptly privatised, grazing pasture was not. Upon the dissolution of co-operatives, individuals were given title to livestock divided among them. Uncertain about their access to pastures, many farmers opted to slaughter their sheep for immediate gain.

Indigenous Kirghiz

Indigenous Entrepreneurship

Natural pastures cover over 9 million hectares of the Kyrgyz Republic. This corresponds to over 85% of the nation's land area. However, this primary natural resource – traditionally the source of livelihood for herdsmen and livestock – has been given insufficient priority. In these rural areas where bears, snow leopards and wolves roam wild, traditional entrepreneurs set up remote yurt encampments. Known in Kirghiz as *beauzi*, a yurt is a round, portable home, made of felt on a wooden frame, and used by nomads of the region. Camels transport the frames. A camel sells for roughly $300 here.

Although Stalin's collectivisation ended traditional nomadic lifestyle, the herdsmen still ride on horseback, tending to their livestock much as their ancestors did centuries ago, and many herders still wear a white felt hat with either black or red trim. Animals and children drink side-by-side from mountain streams. Summer pastures are principally the rolling grasslands on mountain slopes. The changing seasons make migration necessary and winter pastures are areas near the farms where wheat is usually grown. Camels, cattle, goats, horses and sheep are raised for their milk, meat, fibres, skins and pelts; in addition, horses are useful to carry people while camels are good to transport heavier items – such as a yurt, in its entirety. Mare's milk is fermented into *koumis*. Horsemeat stays fresh only three days and therefore horses are slaughtered only for festive occasions. In addition to indigenous black sheep, the wool from which is good for making carpets, white sheep are also kept. The latter have more valuable wool, but the variety is less hardy. Also, white sheep tend to be helpless in the snow and rely on goats to pave the way for them. Dogs are used to round up the animals. Birds of prey are trained to hunt rabbits for their masters.

All of the above are potentially a very valuable resource; however, this is not exploited to its full potential due to lack of knowledge in nutrition, health care and breed improvement. Agriculture must no longer be viewed as a stagnant industry. It is affected not only by changing weather patterns and demographic shifts, but also by changing technology, government intervention, competition and market demands. Currently, in the Kyrgyz Republic, there is a problem of low productivity in the livestock sector. This can be traced to disease, lack of thrift, poor husbandry and underfeeding.

***Koumis* Stand**

Birds of Prey Are Used for Hunting

The Naryn Oblast is a region of the Kyrgyz Republic that has specialised in the raising of animals. The mountain pastures are ideal for cattle breeding. The sheep, here, tend to be fine-fleeced or half fine-fleeced. Horses are also important to the regional economy, and the most profitable branch of stock raising is yak breeding.

Yak

Yaks graze all year long, and do not cost much to keep. Their products have stable demand; this includes hair, meat, skin and wool. When yaks are bred with dairy cattle, hybrids yield twice the quantity of milk. A problem, however, is inbreeding, a result of which, the average weight of yaks has been falling, and fewer young survive. During the 1990s, the average weight of yaks fell by several kilograms.

Over half of the people in the Kyrgyz Republic live in the countryside; among the agricultural and pastoral communities, goats and sheep are an ideal investment. According to unpublished files of the State Commission on Foreign Investments and Economic Assistance, in Bishkek, farming employed two-fifths of the population in 1998, and contributed about 40% of the GNP. The 9.3 million hectares of pastures include 1.2 million hectares that are at an elevation 4,000 metres above sea level – ideal for yaks.

Sixty percent of the gross agriculture production is derived from raising cows and sheep for meat, leather, and wool. In addition to livestock breeding, almonds, apricots, cereals, cotton, cucumbers, currants, figs, grain, grapes, peaches, plums, persimmons, raspberries, tobacco, tomatoes, walnuts, and wheat, grow in abundance in this agrarian country. Also, sugar beets are cultivated in the Chui Valley. The principal region for growing potatoes is the Issyk-Kul Oblast, while cotton and most of the country's tobacco fields are concentrated in the Osh Oblast. It should be noted, however, that only 7% of the land is arable, and draughts occur every four to five years. Water pollution is also a problem, as is the increasing salinity of the soil, a result of faulty irrigation practices. The world's second largest mountain lake has been depleted of fish.

The forage crisis of the 1990s led to massive de-stocking. The population of sheep dropped from 12 million in 1991 to 5 million in 1996. Unjustifiably, there is a lack of technical support to farmers. Nevertheless, meat production in the Issyk-Kul Oblast almost doubled from 1994 to 1995, and that in the Osh Oblast almost quadrupled. During that same time period, eggs and milk production increased four-fold in the Chui Oblast, and almost fourteen-fold in Osh Oblast. In the Chui Oblast, the bazaar prevails.

As a result of the demand for wood, coupled with overgrazing, many mountain pastures no longer have any woody species. Such pastures, which have been depleted, should be re-seeded; the planting of fodder shrubs would be most useful. Also, yak bulls should be brought in, from Tajikistan, for example, to strengthen the Kyrgyz yak population.

Inside the Yurt

Cooking Mutton

Yet, one must not expect a market orientation in the bazaar, as transactions lack a formal framework. As noted by Jaworski & Kohli (1993), there are several antecedents of a market orientation and these are lacking here.

Urban Enterprise

Rapid reform in the Kyrgyz Republic accelerated the creation of new opportunities for formal firms operating in the firm-type sector. This statement merits qualification, however, as the distribution of opportunities is very uneven. International aid and foreign investments have been arriving predominantly in the national capital, Bishkek. It is in this city that change is most noticeable. This, in turn, prompted rapid urbanisation. In Bishkek, an apartment that cost $2,000 in 1993 could sell for more than ten times that amount, just five years later.

It should be emphasised that despite economic growth, only a minority of entrepreneurs prospered as the gap between rich and poor widened. Each morning, the author watched elderly women line up to rummage through garbage containers. In another queue were a blind man, a pensioner and a mentally challenged child each waiting their turn to beg from by-passers. It might be added, also, that foreigners attract swarms of dramatic beggars with seemingly countless offspring.

Capitalism has ushered a variety of imports into the Kyrgyz Republic, and these too are especially visible in Bishkek. Not far from an individual selling carrots, cucumbers, potatoes and radishes laid out on a pavement, is a kiosk selling brandy from Greece, *spumante* wine from Italy and sour cherry juice from Turkey. Other players of the informal economy include an ambulant vendor selling draught beer from his portable tap. A small café nearby sells beer from the Czech Republic, chocolate from Russia, Coca-Cola from the United Arab Emirates, coffee and vermouth from Israel, orange juice from Latvia, pistachios from Italy, tea from England and wine from Germany.

A neighbouring small business sells tomato ketchup imported from Bulgaria, chocolate from France, pineapple juice and vodka from Germany, beer from Israel, wine from Italy and cigarettes as well as spaghetti from the United States. Not far, one can find local vodka, Mitzli brand orange juice from Israel and Dana brand soap from Dubai.

To Market

At the Market

Larger stores sell refined vegetable oil from USAID, labelled "NOT TO BE SOLD OR EXCHANGED." Between a newspaper stand and a watch repair kiosk is an individual selling books at an impromptu stall. Across the street is a table displaying sunglasses imported from China. Also available is shoe polish from China, chocolate from Peru and canned sardines from Thailand. Bazaars are typically open from 6 a.m. to 7 p.m. Smaller markets have no set hours.

"NOT TO BE SOLD OR EXCHANGED" Except for Cash

Grain for Sale

Still No Product Differentiation Across Vendors

During the late 1990s, the elderly received a monthly pension equivalent to approximately $15, and during some months cheques did not arrive. This made it necessary to rely on offspring and/or other sources of income for subsistence. Where possible, families have a small vegetable garden and sometimes a cow along with a couple of sheep, not always practical when living in the city.

Mass migration to Bishkek has also had a negative impact on other areas being depleted of people. In the once populated areas of Jumgal, many houses are boarded up, and the fields are overgrown with weeds. Agricultural machinery is simply rusting. In Min-Kush, the industrial heartland of the Naryn Oblast, mass unemployment has become the norm. There is a serious mis-match between market demand and skills available in the workforce.

The creation of free economic zones may be a solution. Such zones exist in Bishkek, Karakol and Naryn Oblast. Yet, research on their effectiveness is lacking. Future research might focus on evaluating concessions made to foreign investors and entrepreneurs. Future research might also speculate on the establishment of more free economic zones. In the Issyk-Kul Oblast, for instance, produce-processing industry could possibly be developed through the establishment of a free economic zone.

Pluralism

According to the last census conducted in the USSR, the ethnic Kirghiz – of Turkic origin – formed 52.4% of the population of their republic, in 1989. Today, at least 60% of the population in the Kyrgyz Republic is ethnic Kirghiz. There are also 80 ethnic minorities in this country.

While the north of the republic was industrialised, the Russians had less impact on the southern provinces of Jalal-Abad and Osh. People in these southern regions tend to be more conservative and religious than are the northerners.

Although Kirghiz has become the official national language of the independent country, Russian is also recognised as an official language, in provinces where Russians form the majority. About three quarters of the population of the Kyrgyz Republic are Sunni Muslims.

Russians at the Bazaar

Kirghiz Man Wearing Traditional *Ak-Kalpak* Hat

Toward the Future

The social structure of traditional, rural society had its own mechanisms for teaching traditional entrepreneurial values. As well, it had its own intrinsic social web. In contrast, life in the city does not.

As individuals find it increasingly difficult to perpetuate traditional entrepreneurship in the pastures, many move to the city but are not prepared for urban living. They are hoping to stumble upon prosperity, but it is more common to come across invalids and blind people appealing for mercy. Economic despair along with the boredom of unemployment leads some to join the many already involved in alcohol abuse.

Also problematic is energy distribution. Although the Kyrgyz Republic exports electricity, the nation relies on imported fuel, which is often scarce, especially in rural regions.

Another threat to development is an emerging problem concerning ethnic minorities. There appears to be a rise in the level of ethnic tension within the Kyrgyz Republic. Russians perceive themselves as being treated as second-class citizens, while the Uzbeks – who are more traditional and who value cultural identity over economic reform – claim that they are third-class citizens. This situation has resulted in unrest in the south of the Kyrgyz Republic. Escalation in civil strife may cause harm to economic development.

A major concern is the exodus of Russians from the Kyrgyz Republic. In 1989, one out of four residents was Russian; by 2002, the Russian minority was reduced to fewer than one in six.

In the initial phases of modernisation, only a small number of individuals internalise the cognitive and normative themes intrinsic to a modern economy. In the case of Soviet Kirghizstan, it was the Russians who never integrated into indigenous society, but rather introduced industry to the region. More recently, nationalism and independence, followed by rapid reform, have resulted in a new country, where Kirghiz has replaced Russian as the official national language. A minority in this new country, Russians have been feeling threatened and many have emigrated, taking with them their skills.

Informal Economic Activity

Entrepreneurs on the Move

Thus, although policy reform has become a reality, many people are waiting for something to happen. They become increasingly dependent on international aid for subsistence and on foreign firms for products. A major problem is that the shock policy model, which introduced rapid reform here, has not taken into account the psychology of this mellow culture.

The use of culturally insensitive theories, from the West, has contributed to a resentment of what is being called "cultural imperialism." Perhaps then, decision-makers should increase their sensitivity to the culture of this nation and the psychology of its people. Given that the indigenous Kirghiz have been successful herders for centuries, manifesting their entrepreneurial spirit in the countryside, why replace this by urban enterprise? Both rural and urban sectors can be encouraged to evolve.

Kyrgyzstan has repeatedly been the victim of externally imposed changes. The Soviets geared the local industry to service their own military industrial complex, with an emphasis on antimony, mercury and nuclear arms. More recently, rapid reform, for the sake of change, has introduced new problems to the Kyrgyz Republic. New taxes squeeze independent, indigenous entrepreneurs who ask nothing of the state. Meanwhile, Western-style capitalism has been imposed, causing rapid urbanisation and related social problems.

In July 1995, Parliament introduced a new tax on pastureland and another tax on land used to grow hay. Taxing traditional herdsmen should not be the means to reduce state deficits. Rather, these traditional entrepreneurs should be allowed to prosper. Efforts should therefore be focused on teaching the principles of rotational grazing.

Allowing the impoverishment of indigenous entrepreneurs can induce social, cultural and political crises, as well as uncontrolled urbanisation, in addition to the growing pains of the existing economic crisis.

Government deficits are the responsibility of the power structure, and discipline must begin here. The Kyrgyz Republic has been praised for its rapid rate of transition, but social issues have been overlooked. Perhaps transition should not be viewed so much as an end in itself, but rather a means to something beyond.

Chapter 7

The Republic of Tajikistan[11]

Introduction

Tajikistan covers an area of 143,100 square kilometres, making it the smallest of the five formerly Soviet Central Asian republics. It is landlocked – surrounded by Afghanistan, China, the Kyrgyz Republic, and Uzbekistan.

Until 1930, the Tajik language was written using Arabic script. It used the Roman alphabet from 1930 to 1940, and Cyrillic from 1940 to 1992. Today, the Tajik language once again uses Arabic characters.

There have been several changes in government since independence, declared in 1991. Economic development in Tajikistan was delayed by the 1992-1997 civil war, between the government and the Islamic-led United Tajik Opposition. Of the fifteen formerly Soviet republics, Tajikistan has the lowest per capita GDP.

According to the Economist Intelligence Unit, the black market in Tajikistan exceeds 57% of true GDP. In contrast to other transitional economies where small business thrives in the informal sector, Tajik entrepreneurs perceive the covert sector as a most interesting alternative for economic activity.

Tajikistan is the poorest of the five Central Asian republics of the CIS. It is also the least stable.

---

[11] This chapter includes information obtained from: the Ministry of Agriculture; the Ministry of Commerce and Material Resources; the Ministry of Culture and Innovation; the Ministry of Economy and Foreign Economic Relations; and the United Nations Military Observers Team.

Historical Overview

Surrounded today by Turkic neighbours, the Tajiks – people of Persian descent – have been in Central Asia long before the 10th century invasions by the Turkics. Between 1918 and 1924, Tajikistan was part of the Turkestan Soviet Socialist Republic. In 1924 – when the Uzbek Soviet Socialist Republic was carved out of Turkestan – Tajikistan became the Tajik Autonomous Soviet Socialist Republic, within the Uzbek Soviet Socialist Republic. On December 5, 1929, this Tajik region emerged as the Tajik Soviet Socialist Republic – a constituent republic of the USSR. It remained, nevertheless, controlled by Moscow and its capital city, Dushanbe, became known as Stalinabad. Central planning forced out traditional mixed farming – which had enabled self-sufficiency – and instead imposed specialisation in cotton, to the benefit of the Russian economy.

During its years as a constituent of the Soviet Union, the Tajik Soviet Socialist Republic had the highest population growth rate of any republic in the USSR, and the second lowest per capita GDP in the union. The Tajik capital was the base from which the Soviets invaded Afghanistan in 1979.

Occupation by the Soviet army was actually good for the Tajiks inasmuch as it controlled inter-regional tension within the Tajik Soviet Socialist Republic. The same way that India was kept united under the British, or Yugoslavia with Tito, the Soviets maintained peace within each of its republics. Tight control by the Soviets postponed ethnic conflict among Tajiks. When Tajikistan became independent, however, inter-regional tension resulted in civil war and economic collapse. In the economic realm, reforms were meaningless as former communist officials were reluctant to let go of their power.

The former Tajik Soviet Socialist Republic gained its independence, on September 9, 1991, becoming the Republic of Tajikistan – locally known as Jumkhurii Tojikistan. Immediately after independence, *The Economist* correctly predicted a potential for trouble in the region and Tajikistan fell into turmoil, as it experienced several changes of government. Tajikistan tried to adopt the Russian rouble as its own domestic currency, but the instability and unpredictability of Moscow's economic policies resulted in a serious monetary crisis. Workers and pensioners were unpaid for months.

The Islamic Renaissance Party, legalised in 1991, took power in a 1992 coup. Civil war ravaged the country and a state of emergency was imposed in January 1993. Although the United Nations mediated peace talks between the government and its opponents – who agreed to a cease-fire in 1994 – the conflict continued, despite the deployment of Russian troops throughout the country.

Industrial production shrank by 31% in 1994, while the population's rate of growth – in spite of emigration – continued to be 3%, the highest in Central Asia, and four times that in the nearby Republic of Kazakhstan. A new constitution was adopted on November 6, 1994, but a variety of factors prevented any significant reform, a prerequisite to the establishment of an entrepreneurial class. The Tajikistani rouble replaced Russian currency on May 10, 1995, making Tajikistan the last country outside Russia to abandon the Russian rouble.

Tajikistani Roubles

By 1996, increasing tension and the continuation of shootings strained the environment for any form of legitimate entrepreneurial activity. A cease-fire was agreed to on December 23, 1996. After the June 27, 1997 peace agreement was signed, 5,000 fighters of the opposition were integrated into the Tajik army. Yet, fighting resumed in November 1998.

On February 27, 2000, Tajikistan held is first multi-party parliamentary elections. However, international monitors reported that it fell short of minimum standards. As the national currency rapidly lost its value, the somoni was introduced on October 30, 2000. The new unit was defined as worth 1,000 Tajikistani roubles. Not surprisingly, the economy has a significant covert sector.

Covert Economic Activity

Policy & Transition

Transition to a market economy, in Tajikistan, has been delayed by civil war. The combined effect of fighting, the loss of Soviet subsidies and the breakdown of Soviet distribution channels, resulted in the collapse of the Tajik economy. The nation has been relying on humanitarian assistance for subsistence. Uzbekistan has been a major contributor, but fearing a spill-over of civil unrest, religious fervour and military conflict, this country has often closed its border with Tajikistan, and has disrupted fuel supplies as well.

This economic situation in Tajikistan is ironic, given that the republic is particularly endowed with a variety of natural resources, including antimony, coal, lead, mercury, petroleum, tungsten, uranium and zinc. In addition, it has significant potential for hydroelectric power; national efforts have been directed toward this option, but not in a way to maximise its potential. An aluminium plant, some obsolete small-scale factories involved in food processing, and other light industry are among the few signs of formal business activity in this country.

Increasing pressure is being placed on this nation, as the population exceeds 6 million people, with a labour force of about 2 million. The nation has failed to become self-sufficient and a low standard of living persists. This problem is worsened by the fact that women continue to bear an average of five children – compared to three in the neighbouring Kyrgyz Republic – and there is no system set up for the youth to grow into. A remarkable 43% of Tajikistan's population is under 15 years old; this is 6% more than is the case in the Kyrgyz Republic, and 12% more than the figures for Kazakhstan.

Numerous constraints in Tajikistan discourage legal forms of entrepreneurial activity. Continued hostilities between ethnic groups contribute to a variety of problems, including strains on the environment, infrastructure inadequacies and the lack of importance given to basic education. Among the relevant issues are: over-utilisation of water, increasing pollution, excessive use of pesticides, inadequately developed and poorly maintained communication networks, chronic fuel shortages, worsening international relations, border conflicts and low attendance at school. In contrast, covert economic activities are quite promising due to their accessibility and uncontrolled channels of distribution. The lack of trust in the unstable currency has allowed the barter system to thrive over a modern cash economy. Cognitive innovation is necessary, as payments in kind require a different notion of measurability than do cash payments. Different elements enter the equation and these change social relations.

Barter at the Bazaar

Some entrepreneurs see tourism as a fast dollar sector. However, it should be noted that in contrast to neighbouring Uzbekistan, which benefits from tourists visiting ancient centres of the Silk Road network, Tajikistan has almost no tourism. Concerns include inadequate sanitary facilities as well as security. Fuel shortages aggravate the situation, as people coming into the country are unsure about if and when they can get out. Nevertheless, a very big attraction in Tajikistan is the world's largest population of wild Marco Polo sheep. These animals have horns up to six feet long, and hunters from Germany and the United States have been paying $18,000 for a kill. A problem, here, is that experts estimate that there are only 8,000 adult Marco Polo sheep left.

Despite the so-called peace, the environment is still perceived as risky for investments. Instead, Tajikistan has become a major centre for the distribution of illicit drugs to Europe and to North America.

Corruption is the norm. Policy-makers – riding in fancy limousines – claim that they lack the funds to chlorinate the water supply of the republic.

Agriculture

The economy in Tajikistan is largely agricultural; during the late 20th century, half of the labour force worked on farms, producing cotton, fruits and vegetables. Grapevines are grown not only for the grapes but also for the leaves, which are used in a variety of dishes. Farming is very labour-intensive; one can observe literally hundreds of women hunched over in fields.

The agricultural sector has severe limitations, as only 6% of the land is arable. Tajikistan is part of the basin of the Aral Sea, and this region is suffering from severe over-utilisation of water; increasing levels of soil salinity are a problem. Furthermore, industrial pollution and the excessive use of pesticides have contaminated much soil. Cattle, goats and sheep are raised, but meadow and pasture cover less than one-quarter of the land, and animals are forced to graze on the shoulders of motorways. The country has no forest and the Pamir Mountains have semi-arid to polar weather.

Cotton is officially the number one crop, but only because the former USSR imposed its production. In fact, the land in Tajikistan is not ideal for this labour-intensive crop. Furthermore, the development of secondary industries – such as textiles – has not been an attractive option. Instead, there has been a significant rise in the cultivation of cannabis and opium poppies, both of which are less labour-intensive and much more profitable than is the case with cotton. In the north-west of the republic, near Lake Dushakha, heroin poppies appear to be the principal cash crop.

Infrastructure

Enterprise is hindered by the general lack of infrastructure. Communications are poorly developed and the system is badly maintained. Fewer than one in ten persons in Tajikistan has a telephone, and entire towns are excluded from the national network. This can be a difficult barrier when conducting business.

Also plagued is the transportation system. The entire country has only 480 kilometres – fewer than 300 miles – of rail line in common carrier service and bus service is practically non-existent. During the summer, motor traffic is disrupted by fuel shortages, and during the winter, roads are blocked by snow and seldom cleared. There are few gas stations; along major highways, drivers purchase gasoline in bottles, from entrepreneurs who operate at the micro-enterprise level, but this is only when fuel is available.

No Fuel, But Plenty of Alcohol

The geographic location of Tajikistan could be an important asset in itself. In neighbouring Uzbekistan, a snow-covered mountain range separates the capital city, Tashkent, from Kokand – in the Fergana Valley – and points north; therefore, Uzbek buses link towns in Uzbekistan, via Tajikistan. This provides transportation within Tajikistan, and business to Tajik entrepreneurs along the way. However, even after the 1994 cease-fire, border closings "for security reasons" were disruptive to business. Fear of being knifed or hurt from homemade bombs are real concerns for the few people who by chance wish to pass through. The town of Hudzand, alone, has the potential of making Tajikistan an international silk centre. However, entrepreneurs engaged in import/export operations are concerned about border closings and international disputes. Tajikistan's boundaries – both with China and with the Kyrgyz Republic – are in dispute. In 1996, the border with Afghanistan was the site of numerous killings. Better relations with one's neighbours might be more conducive to cross-border entrepreneurial activity.

As for inter-continental links, in 1994, Tajik Air leased a Boeing 747 from United Air Lines, and introduced air service between Europe and Tajikistan. This might have helped entrepreneurs in the short run, but the choice of aircraft was not a wise one, as the 747 – even with a low payload – consumed much jet fuel while Tajikistan was experiencing fuel shortages. Moreover, the passenger capacity of this particular aircraft was much higher than demand for seats. The national airline finally collapsed. More recently, Tajikistan International Airlines launched service as far west as London's Heathrow Airport; British Airways agreed to be the handing agent.

Education

Despite logistic difficulties, an exchange programme with a management school in Nebraska has allowed students at the management faculty of Hudzand State University, in Tajikistan, to benefit from US-trained professors. Courses include finance and marketing, but reportedly emphasise working with a large-scale corporation. Since few large firms are set up in Tajikistan, perhaps more emphasis should be placed on entrepreneurial skills and on self-employment. In addition, such training should be made available to a wider audience. Individuals are currently engaged in micro-enterprise; at Kanibadan bus station, for example, children sell tea to passengers – but such vendors of the informal sector are unlikely to attend a five-year programme at a business faculty.

Why Study?

In fact, education is generally a problem altogether. Whereas Soviet rule had imposed universal free education, the situation changed during the late 20th century, when a small percentage of Tajik children were attending school.

Most Tajiks speak Tajik, which has become the new official language; however, Russian is still necessary for certain government matters. This is problematic for the rural Tajik who has not learned Russian. There is thus a constant catch-22 situation in Tajikistan, and the economic environment presents little opportunity for legitimate entrepreneurship.

Pluralism

In contrast to the other peoples of Central Asia who are of Altaic stock, the Tajiks are racially Indo-European, and Mediterranean features are common. While the other republics of the region speak Turkic languages, the Tajik tongue is related to Farsi. A difference with possibly greater implications is that while Kazakhstan, Turkmenistan, Uzbekistan and the Kyrgyz Republic have been looking at Turkey as a role model, Tajikistan has leaned more towards Iran. Whereas Turkey has been striving to improve its economic well-being, Iran concentrated first on strengthening religious ideals. This is manifested by such acts such as the funding, by Iran, of the major mosque in Dushanbe. However, emphasis swayed away from religion, in May 1997, when the Iranians elected Mohammed Khatami, a moderate leader.

At the time of the 1989 census – the last conducted in the USSR – 62.3% of the population of Tajikistan was Tajik, 23.5% Uzbek, and 7.6% Russian. The Tajiks are predominantly Sunni Muslims. Within Tajikistan is the Badakhshan Autonomous Province, whose inhabitants are predominantly of the Ismaili sect of Shi'ite Islam – with no mosques. These people claim to be descendants of people who were subjects of Alexander the Great. Their sect is led by the Aga Khan.

Lately, many of the minorities in Tajikistan – including the Russians – have been feeling insecure, and some have moved away, contributing to the country's net emigration rate. As an illustration, membership at the Dushanbe synagogue fell from 20,000 in 1989 to 10,000 in four years.

Uzbek Gives His Back to Tajiks

Tajikistan today is the victim of the historically-rooted regional animosities which climaxed into serious civil war as people from the Hissar, Kulyab and Leninabad regions of this country, have been resented by others who have felt relatively deprived in the political realm. Young (1971) suggested that entrepreneurship occurs when an ethnic group with low status is denied access to mainstream society, but only when resources are available. In Tajikistan, the politically deprived are also economically deprived.

Toward the Future

Tajikistan is pluralistic, with a shared core universe, and different partial universes coexisting with the lack of mutual accommodation. The perpetuation of hostilities, despite the official cease-fire, has worsened the general economic situation. Kazakhstan, the Kyrgyz Republic, Russia and Uzbekistan have accepted to defend the Afghan-Tajik border. Yet, ethnic violence from within the country has threatened the sustainable existence of the Tajik people. Professionally trained mercenaries from Afghanistan, Iran and Pakistan were among those fighting inside Tajikistan.

The mountain region of Gorno-Badakshan is an Islamic rebel bastion within Tajikistan, but beyond the control of the Tajik government. Anarchy reigns here, contributing to the thriving of opium merchants who, from here, travel west into Europe and beyond. The road from Khorog, at the Afghan-Tajik border, is among many routes travelled by hardy nomads transporting illicit cargoes. Border guards are seemingly turning a blind eye to gift-bearing "merchants" engaged in such covert entrepreneurial activities.

Toward a Market Economy

The move to a market economy is process-driven. It requires people to understand the dynamics of a free enterprise. However, due to considerable communist influence in the past, many Tajiks are not familiar with "normal" business practices in a market economy. The Tajik concept of enterprise is not the same as that in the West, and management is not seen the same way. Under the old regime, there were no managers in the Western sense, but rather highly influential industrial bureaucrats. They did not manage; instead, they received orders from superior functionaries and simply executed the instructions. Consequently, people are still waiting for directives. Many are waiting for foreigners to initiate new ventures. Meanwhile, engaging in covert activities is providing an overwhelming source of hard currency, this to the detriment of engaging in legal entrepreneurial behaviour such as that found in internal, informal and formal categories of economic activity.

Whereas independence from Moscow has enabled entrepreneurs in some newly independent countries to enter better economic times, for Tajikistan, secession has not contributed to legitimate entrepreneurship. Instead, Tajiks perceive *perestroika* as having caused a loss of job security, and independence has resulted in a loss of subsidies and distribution channels. Furthermore, the departure of the Soviets helped rekindle ethnic and regional tensions. Aside from trading illegal drugs and/or hunting endangered species, Tajiks see little opportunity for entrepreneurial activity in their new country.

Informal Economic Activity

Chapter 8

The Republic of Turkmenistan[12]

Introduction

The Republic of Turkmenistan covers an area of 488,100 square kilometres, bordering Afghanistan, Iran, Kazakhstan, Uzbekistan, and the Caspian Sea. Turkmenistan is rich in energy, with three trillion cubic metres of natural gas – the fourth largest reserve in the world, after Russia, the United States and Iran. In addition, Turkmenistan has 700 million tonnes of oil reserves, and the world's third largest sulphur deposits. As well, this country is among the ten largest cotton producers in the world. Despite these riches, the people of Turkmenistan have remained generally poor. Under Soviet rule, poverty was blamed on the fact that the republic was required to sell its gas and cotton to Russia at artificially low prices.

Even after independence, the environment for enterprise in Turkmenistan has not been conducive for the creation of widespread prosperity. O'Driscoll, Holmes & Kirkpatrick found more government intervention in this country, than anywhere else in Central Asia; they noted, "Corruption is a major impediment, and the Economist Intelligence Unit reports that any thorough reform is unlikely for political reasons (2001, p. 368)."

---

[12] This chapter includes information obtained from: the Central Bank; the Department of State Property and Privatisation; the Ministry of Economy and Finance; the Ministry of Foreign Economic Relations; the Ministry of Industry; the Ministry of Trade; the Small and Medium Enterprise Development Agency; the Turkmenbank; and Turkmenavia Airlines.

Although Turkmenistan is largely covered by desert, agriculture provides a significant proportion of the nation's employment opportunities, accounting for almost half of national GDP. Much agriculture relies on irrigation.

Much of the Country is Desert

Historical Overview

Alexander the Great established a city here, on his way to India. The Orguz Turks, forefathers of the Turkmen people are said to have arrived here during the late 10th and early 11th century. It is believed that the early Turkmen people were nomadic horse-breeding clans.

The Russian Empire gradually expanded into Turkmen lands, between 1865 and 1885. By 1894, the Tsar ruled all of Trans-Caspia from Moscow.

In October 1917, the Tashkent Soviet assumed authority of Turkestan. The quasi-independent khanates of Bokhara and Khiva were transformed into republics, in 1920. On October 27, 1924 the Trans-Caspian Region of Turkestan Autonomous Soviet Socialist Republic was fused with parts of Bokhara and Khiva to form the Turkmen Soviet Socialist Republic, and this entity was recognised as a constituent republic of the USSR on May 25, 1925.

Central planning was greatly feared and many Turkmen fled to Afghanistan to escape collectivisation. During the 1930s, cotton became the backbone of the Turkmen economy. Despite bumper crops, the republic did not prosper, as the cotton was sold to Moscow at prices below world market values. In the absence of processing facilities, Turkmenistan was left out of the value-adding phases. Cotton was exported at low prices, and cloth was imported at prices dictated by central planners.

Sovietisation also involved the decline of religion. Of the 441 mosques that stood in Turkmenistan in 1911, only five were still standing in 1941.

Prospects brightened only when Turkmenistan became independent on October 27, 1991.

The Turkmen Model of Limited Privatisation

Under central planning from Moscow, the Turkmen economy was focused on cotton. Since the economy has changed paths, oil has become increasingly important. A new constitution was adopted on May 18, 1992. Joint ventures became a popular means to combine capitalist ideas with local enterprise. In 1993, the Daykhanbank – a local state-owned bank specialising in agricultural projects – joined forces with the largest bank in Turkey – the state-owned Ziraat Bank – to create the Turkmen-Turkish Joint Stock Commercial Bank.

At Home

The Toilet

Cotton

At the Bazaar

Turkmenistan launched a cautious programme of limited privatisation on June 1, 1994; also that year, the manat was devalued. In May 1995, a railroad was opened between Iran and Turkmenistan; this provided Central Asia with a link to the Indian Ocean, cutting travel time between Europe and Southeast Asia by several days. In December 1995, President Niyazov initiated the President's Programme for Social and Economic Development in Turkmenistan. As recommended by the IMF, key objectives of this programme included the implementation of market reforms, making the manat internally convertible and restructuring the economy with a focus on controlling the expansion of credit. The government subsequently announced more ambitious privatisation plans, and the service sector was privatised.

Presidential decrees have the power to restructure joint stock commercial banks into state banks, and the banking sector was restructured during the late 20th century. In December 1998, three major banks were declared state banks with narrowly defined scopes of activities. In January 1999, the number of banks operating in Turkmenistan was reduced from 67 to 13. Fifty-two peasant commercial banks became branches of the Daykhanbank. Other state banks include the State Commercial Investbank, the Turkmenbank and the Vnesheconombank – the State Bank for Foreign Economic Affairs of Turkmenistan – that was given the monopoly for all foreign exchange operations involving the government. State firms in Turkmenistan have been ordered to conduct business only with state banks.

Entrepreneurship

Turkmenistan approached entrepreneurship issues with caution. It allowed the creation of 6,000 small-scale firms in 1992, but delayed auction sales of mid-size and larger businesses.

In 1993, the Cabinet of Ministers founded a joint venture with the Khal Bank of Turkey, two state-owned Turkmen banks, and private firms in Turkmenistan. The result was the International Joint Stock Bank for Reconstruction, Development and Promotion of Entrepreneurship. However, individuals interviewed by the author expressed doubts as to how effectively entrepreneurship was being promoted. It appears, nevertheless, that the Rossiysky Credit Bank – Russian-Turkmen Joint Stock Bank – provides consulting services that assist entrepreneurs.

Soap

Safron

The Rossiysky Credit Bank also offers a matchmaking service, providing opportunities for export. Interviewees claim, however, that opportunities were curtailed when recent legislation restricted access to foreign exchange, to certain persons. Respondents also told the author that high inflation rendered business plans useless. According to O'Driscoll, Holmes & Kirkpatrick (2001), annual inflation averaged 77.41% between 1991 and 1999.

Since 1995, the Small and Medium Enterprise Development Agency – a joint venture between the European Union and the Government of Turkmenistan – has been providing entrepreneurs with information on business law, taxation and related matters.

A first glance might suggest that entrepreneurship is facilitated in Turkmenistan; in actuality, entrepreneurs in Turkmenistan face considerable regulation. Article 1 of the Law on Entrepreneurial Activity provides a definition: "Entrepreneurship is initiative, independent activity of citizens aimed at obtaining profit or income and carried out on one's own behalf, at one's own risk and on one's own property liability of legal entity – enterprise." Article 3 specifies that it "is prohibited to carry out entrepreneurial activity without registration." Article 9 states that an individual may apply for registration of a sole proprietorship – without creating a legal entity – and this be done by local executive bodies at the place of activity of an entrepreneur in the order determined by the Cabinet of Ministers. Registration may be refused, in which case the law allows appeal, in court. However, of all the individuals interviewed by the author, appeals seem to have helped none. When registration of an entrepreneur is permitted without the creation of a legal entity, the sole proprietor cannot hire labour. In order to have employees, a firm must be registered – as detailed by Article 26 of the Law of Turkmenistan on Enterprises – with Commissions formed at khyakimliks of velayats and Ashgabat. Several businessmen complained to the author about the high cost of registration.

Article 9 also rules on the registration of entrepreneurial activity with the establishment of a legal entity, in accordance with the elaborate Law of Turkmenistan on Enterprises. Section III of the Law on Entrepreneurial Activity elaborates on regulations pertaining to entrepreneurs.

The Law of Turkmenistan on Enterprises is equally detailed. Article 17 of this law states that the list of output – for sale at state prices – is determined by the Cabinet of Ministers of Turkmenistan, by agreement with an enterprise. An informer told the author, however, that an enterprise might not really have the option to disagree.

Article 21 of the Law of Turkmenistan on Enterprises states that an enterprise is obliged to take action to improve labour conditions. One interviewee told the author, “In order to be a successful entrepreneur you must first be a lawyer, or able to buy the law.”

Most of the self-employed, in this country, operate micro-enterprises, and these are often informal. Street vending is common. Vendors – dressed in colourful scarves and mismatched clothing – line the sides of dirt roads in hopes of selling their fruits and vegetables. Mothers supervise their children with regards to techniques. Each vendor has a couple of colourful plastic pails filled with produce. They offer, for sale, cherries, onions, peppers, plums, potatoes, and tomatoes, depending on what is in season. As potential buyers approach, sellers inch closer to listen to the sales techniques, and anticipate the outcome.

Pluralism

Seventy-seven percent of the residents of Turkmenistan are ethnic Turkmen, and the country ranks among the more homogeneous of the republics of Central Asia. The Uzbek minority accounts for 9% of the population. In contrast to Kazakhstan, a country that was greatly Russified under the Soviet regime, Turkmenistan was the least Russified of the Central Asian republics. Less than 3% of the people in Turkmenistan are ethnic Russians. To avoid ethnic tensions, Article 28 of constitution forbids the creation of political parties along ethnic or religious lines.

Most (87%) Turkmen nationals are Sunni Muslims. About half the population lives in urban areas.

Toward the Future

Turkmenistan is said to have great potential. It is rich in fuel, and its weather allows for a long growing season. Its cotton is considered among the best in the world. Furthermore, its location puts it at the crossroads of Europe and Southeast Asia. Also in its favour is the fact that the Turkmenistan Internet domain “.tm” is very popular around the world, and this provides significant export revenues – an uncommon export!

Nevertheless, enterprise is crippled here. Heavy paperwork requirements are a strain to owner-managers with limited resources. The multitude of local, profit, and property taxes further deters legitimate entrepreneurship. Frequently changing regulations and tax laws limit entrepreneurial expansion. As concluded by O'Driscoll, Holmes & Kirkpatrick, "The legal reforms necessary for development of a market economy have not been put in place (2001, p. 368). "

Traditional Enterprise

Cinnamon

Toward the Future

Chapter 9

The Republic of Uzbekistan[13]

Introduction

The Republic of Uzbekistan covers 447,400 square kilometres, bordering Afghanistan, Kazakhstan, the Kyrgyz Republic, Tajikistan, Turkmenistan, and the Aral Sea. In January 2000, Uzbek border guards entered Kazakhstan and unilaterally extended their country by marking out a 60-kilometre stretch. Uzbekistan has the largest open-pit gold mine, and the fourth largest gold reserves, in the world. Cotton accounts for 80% of the nation's exports.

The Uzbek language used Arabic script until 1929, and the Roman alphabet from 1929 until 1940 – when Stalin imposed Cyrillic. In 1994, the state began to phase out the use of Cyrillic.

Rather than precipitate transition, this republic has opted for a gradual approach to economic reform. Independence led to an Islamic revival. Tension, in 2001, led to the closure of about 900 mosques. Uzbekistan received $100 million for its support of the United States in the War Against Terrorism.

Defining economic freedom as "the absence of government coercion or constraint (Beach & O'Driscoll, 2001, pp. 43–44)", O'Driscoll, Holmes & Kirkpatrick (2001), reported that Uzbekistan repressed economic freedom; the same study found the government in Uzbekistan to have greater fiscal burden than any of the following: China, Kazakhstan, the Kyrgyz Republic, Tajikistan, Turkmenistan, Myanmar, Cambodia, Laos and Vietnam.

[13] This chapter includes information obtained from: the Ministry of Agriculture; the Ministry of Finance and Economy; the Ministry of Foreign Affairs; the Ministry of Foreign Economic Relations; the Ministry of Industry, Fuel, and Power; and the Ministry of State Property and Privatisation.

Historical Overview

Waves of nomads came to this region 2,000 years ago. By the 2nd century, Samarkand was a wealthy trading centre, prospering thanks to its location at the junction of caravan routes. During the 6th century, nomadic Turks settled here. Arabs brought Islam during the 8th century. By the 10th century, Bukhara – situated along the silk traders' route between China and Rome – was thriving as the capital of a feudal state that stretched across Central Asia.

Genghis Khan conquered Samarkand in 1221. The Mongol conqueror, Tamerlane the Great took the oasis settlement in 1369, at which time it became his imperial capital. Tamerlane became a local hero, as he invaded Afghanistan and Persia in 1381, Russia in 1391, Georgia and Iraq in 1392, India in 1398, and Anatolia and Syria in 1404. He died in 1405, on his way to China. His dynasty lasted until the 16th century, when Samarkand was controlled by a new Turco-Mongol people – the Uzbeks.

Samarkand

The Russians became interested in the area during the early 18th century, when the Khan of Khiva approached Peter the Great and offered to become the vassal of Peter – in exchange for protection from Kazakhs and Turkmen. Khiva was already a prosperous commercial centre, as trade from China, Mongolia and Persia, passed through Khiva on the way to the Baltic Sea and the Black Sea. In 1717, the Russians arrived in Khiva and they soon bought cotton and slaves in Bukhara.

In 1868, the Russian Empire absorbed Bukhara, but the emir survived. In 1920, the Bukhara People's Republic was formed. The Uzbek Soviet Socialist Republic was created on October 27, 1924, when decision-makers in Moscow artificially united the remains of three 19th century khanates.

During WWII, Stalin feared that Crimean Tatars, ethnic Koreans, and Volga Germans in the USSR might be pro-Axis. He therefore resettled large numbers of these people to the Uzbek Soviet Socialist Republic.

Proud of His Medals

After the war, the Russians imposed cotton planting on the Uzbeks, replacing traditional mixed farming. This was a shortsighted strategy, because cotton needed water. The Soviets set up an irrigation system, but this caused the Aral Sea to shrink by half its area. As its shores receded, the Uzbek port of Moynaq, home to 2,000 people, found itself some 40 kilometres from water. In 1978, the Uzbek Soviet Socialist Republic adopted its constitution, also known as the Fundamental Law.

Uzbekistan became independent on August 31, 1991. Independence Day is celebrated on September 1, in accordance with Article 77 of the Code of Labour Laws of the Republic of Uzbekistan. Years after independence, the country still displays scars of the Soviet regime. The port of Moynaq is still 40 kilometres from the sea.

After Independence

The Uzbek Model of Gradual Transition

During Soviet rule, almost 24,000 mosques in Uzbekistan were closed. The observance of Ramadan was banned, as was the wearing of veils, and the sales of the Koran. While independence from the USSR prompted people in other formerly Soviet republics to focus their efforts on the economic realm, millions of Uzbeks viewed independence as an opportunity to return to religion. In 1991, mosques were being opened across Uzbekistan, at the rate of ten per day. With this religious revival, polygamy has been on the rise.

In contrast to the countries that rushed into market economies, Uzbekistan established policies opposed to the shock policy approach. The Uzbekistan Communist Party transformed itself into the Popular Democratic Party – led by President Karimov, former leader of the Communist Party – and with the same people in power, reform was slow. The government sought to maintain a Soviet-style economy with subsidies, price controls, and centrally planned production. This buffered the economy from the sharp falls in output, experienced elsewhere. However, export taxes complicated international trade. In 1991, the Enterprise Tax Law was issued. It was harsh and subsequently amended in 1992 and thereafter.

On December 8, 1992, the 1978 constitution was deemed to have lost effect, as a new constitution was adopted, at the Eleventh Session of the Supreme Council of the Republic of Uzbekistan Twelfth Convocation. The Supreme Council then decreed that December 8 would become a national holiday, Constitution Day. Chapter 1, Article 4 of the 1992 constitution clearly states that while the one state language of the Republic of Uzbekistan shall be Uzbek, the state "shall ensure a respectful attitude toward the languages, customs and traditions of all nationalities and ethnic groups living on its territory, and create the conditions necessary for their development." Chapter 6 of the constitution includes every citizen of the Republic of Karakalpakstan as simultaneously a citizen of Uzbekistan. Chapter 9 guarantees the right to own property (Article 36) and the right to choose an occupation (Article 37).

Chapter 17 is dedicated to the Republic of Karakalpakstan, a "sovereign republic" (Article 70), with its own constitution, provided that it is in accordance with that of Uzbekistan (Article 71).

Until 1993, Uzbekistan had a conservative privatisation programme. The state sold primarily non-transferable shares, to employees only, and at a fixed price.

In 1993, Uzbekistan was expelled from the Russian rouble zone. On November 10, 1993, provisional sum-coupons were issued, and these circulated along with Russian roubles until January 31, 1994, when the coupon became the sole legal tender currency in Uzbekistan.

A vast parallel economy arose. By May 1994, $1 bought 3,300 sum-coupons at the bank, but people in the streets were offering 25,000. That month, the government passed the Law on Foreign Investment, legislation that provides for unspecified compensation in the event of expropriation.

In July 1994, the sum was introduced, equal to 1,000 sum-coupons. The sum-coupon was then withdrawn on August 1, 1994. The new currency experienced rapid devaluation. By October 1994, banks were giving 22 sums per dollar, while the parallel economy offered 33.

Uzbek Currency

Until 1995, enterprises were required to pay taxes based on sales. This was problematic, however, as sales were not necessarily proportional to profits. In April 1995, a tax on profits was introduced. Coca-Cola was also introduced that year. In November 1995, 175 business people, from around the world, attended a two-day conference at the Metropole Hotel, in London. The theme was "Doing Business in Uzbekistan," and delegates were briefed on Uzbekistan's import substitution strategy, for grain and oil in particular.

On January 1, 1996, the republic tightened its visa policies. By May, banks were trading the dollar at 37.5 sums. The author recalls being offered 52, by parallel traders.

President Karimov continued to pursue his model of gradual transition. Citing the Uzbek principle "Don't destroy the house before you have built another one," he noted that shock therapy in Russia discredited the market economy. His message was, "We do not reform things for the sake of reform," and he argued that the speeding-up of reform caused the impoverishment of the people.

Nevertheless, by 1996, prices were liberalised and the 1990s in Uzbekistan were characterised by pains of transition – albeit slow. The monthly salary of a university professor was equal to the cost of a single tank of petrol in the parallel economy.

Due to disputes with Kazakhstan, Uzbekistan frequently cut off the flow of gas to its neighbours. In turn, Kazakhstan disrupted telephone service.

In 2002, farmers were still told what to grow, and their harvest continued to be taken by the state. For them, little had changed since the demise of the USSR.

Little Change for this Man

Uzbek Farmer

Coming to the Market

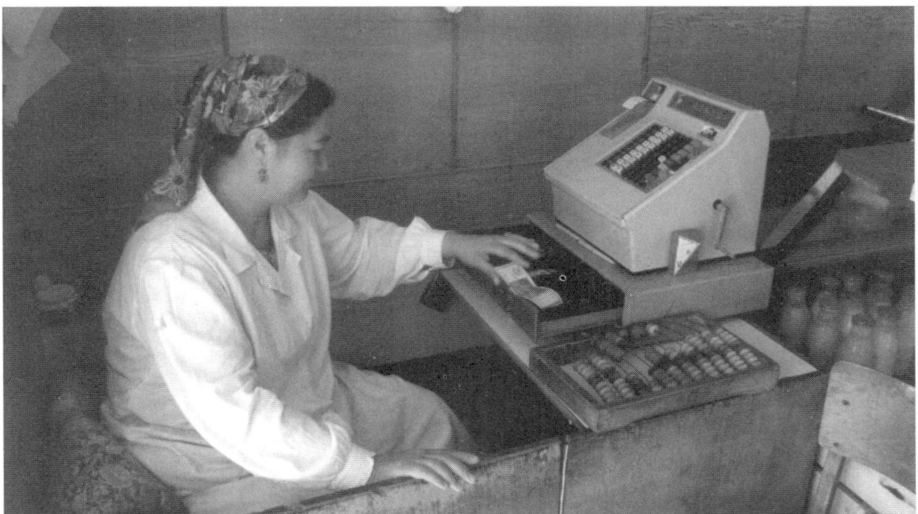

Abacus and Cash Register Used Simultaneously

Entrepreneurship

In January 1994, the Uzbek president issued an edict "On Measures to Further Deepen Economic Reform, Facilitate the Protection of Private Ownership and Develop Entrepreneurship." Yet, entrepreneurship in Uzbekistan is seldom Schumpeterian (1912; 1928; 1934; 1939; 1942; 1947; 1949) in nature. Innovation is limited. Although the Uzbek Soviet Socialist Republic produced excellent engineers, hundreds of these have moved away and resettled in Israel.

Conditions in Uzbekistan give rise to entrepreneurs in the Kirznerian (1973; 1979; 1982; 1985) sense. Individuals identify opportunities for arbitrage, and taking risks as described by Cantillon (1755) and Knight (1921). Often, self-employment is not a desired activity, but merely a means to survival. For many, it is a part-time effort, required to supplement a low salary. According to the November 10, 2001 issue of *The Economist,* doctors in Uzbekistan were earning the equivalent of $15 per month at the time (p. 42).

Pursuing Opportunities

In the lobby of the Chorsu Hotel – where the price of a room includes a calf's tongue for breakfast, but no toilet paper – entrepreneurs of the parallel economy offer to buy dollars a rate superior to the "official black market rate." It turns out that these buyers have a full-time job, and this is a sideline.

A woman boards an inter-city bus to sell pastries, bread, and tea, to passengers; an elderly man signals that he would like one slice of cake. It is easier for him than disembarking to purchase plov – a rice dish with mutton, and carrots with or without turnips – from a local hawker. The woman also has a full-time job, but she says it pays her less than the proceeds of her part-time micro-enterprise. When the bus prepares to depart, her prices fall, reflecting more supply than demand.

As was the case during the zenith of the Silk Road, the bazaar is thriving with social as well as mercantile activities. A merchant guides his donkey-drawn cart past locals drinking tea under hanging sheep carcasses. He sneers at drunken Russians as he proceeds to his market stall.

Informal Stall

At the Bazaar

No Differentiation

At the market, prices often rise as the day progresses. This reflects decreasing supply. For sale are carpets and garments, fruits and vegetables. There are cans of Coke from Korea, Sprite from Israel, and 7-Up from Turkey. Sheep's gut is sold by the metre. Cigarettes are sold one by one. Tu-134 brand of cigarettes seem popular, perhaps because the name is the same as that of a commercial jet.

Typical of the bazaar, radish-dealers are sitting side-by-side. Segmentation refers to the clustering of sellers. There is no brand differentiation. The same is true of the women selling nuts and dry fruit – caravan food, as was sold centuries ago.

Not far, shoe repairmen are sitting side-by-side on the pavement. Again, there is no brand differentiation. Each has his own clientele. Purchases are made according to personal connections. The focus is not so much on the product or service, but on relationships.

Relationships

Pluralism

While the Uzbeks are the largest Turkic group outside Turkey, their country is ethnically diverse. Eight out of ten people here are ethnic Uzbeks. The others include remnants of the German, Korean and Tatar communities, as well as Kazakhs, Russians, Tajiks, and smaller ethnic groups. In addition, there are those who identify themselves as Karakalpak nationals.

Muslim men, wearing traditional black pillbox hats, watch the Russian girls in short skirts. Several languages are heard in the streets and in the markets. The Uzbek language includes Arabic, Persian and Turkish words.

Although many Uzbeks regard themselves to be superior to the Kazakhs and the Kyrgyz, there appears to be more tolerance here than in neighbouring Tajikistan. There has been tension, however, as fundamentalist Wahhabi Muslims – originally from Saudi Arabia – have been demanding that sharia religious law be enforced in this primarily Sunni country.

Minority in Uzbekistan: Jewess

Toward the Future

Uzbekistan is a land of paradox. There are over 250 airports, but most have unpaved runways. When it took 14 hours to process one incoming flight at Tashkent Airport, the author witnessed the airport restaurant serving as a dormitory for passengers waiting for immigration officials to arrive. Indeed, Soviet bureaucracy survived the demise of the USSR.

In contrast to Kazakhstan and the Kyrgyz Republic, Uzbekistan proceeded slowly with privatisation and transition to a market economy. At the time of writing, in 2002, the economy was crippled as potential investors are discouraged by restrictions on currency movements.

Transition to a Market Economy

Chapter 10

The Union of Myanmar (Burma)[14]

Introduction

Myanmar – formerly the Union of Burma – covers 678,500 square kilometres, bounded by Bangladesh, China, India, Laos, Thailand, the Bay of Bengal and the Andaman Sea. In approximately 1295, Marco Polo described the wonders of Burma with astonishment (Moore, 1963). In 1898, Kipling said of Burma, "it will be quite unlike any land you know." Garrett used the words "virtually unknown to the outside world (1971, p. 343)." Hodgson summarised Burma as the "most reclusive country of mainland Southeast Asia (1984, p. 90)." Swerdlow described Myanmar as "the secretive nation (1995, p. 73)." Hirsh & Moreau called it "East Asia's last frontier (1995, p. 10)."

Hagen noted, "the Burmese level of living was higher than that of India or China (1962, p. 436)." Until the 1950s, Burma was the world's foremost rice exporter. After an era of socialism, a market-oriented economy was introduced in 1988. Since then, many of Myanmar's social indicators – including literacy rate, and doctor/population ratio – have compared favourably with those of its richer neighbours (Thein, 1996). Yet – although the country is generously endowed with hardwood forests, oil, precious gems, silver and other minerals – its people have not prospered.

---

[14] This chapter includes information obtained from: the Ministry of Agriculture and Irrigation; the Ministry of Co-operatives; the Ministry of Commerce; the Ministry of Industry; the Ministry of Information; the Ministry of National Planning and Development; the Ministry of Rail Transportation; the Myanmar Timber Enterprise; the Union of Myanmar Chamber of Commerce and Industry; and the University for the Development of the National Races of the Union.

Traditional Home

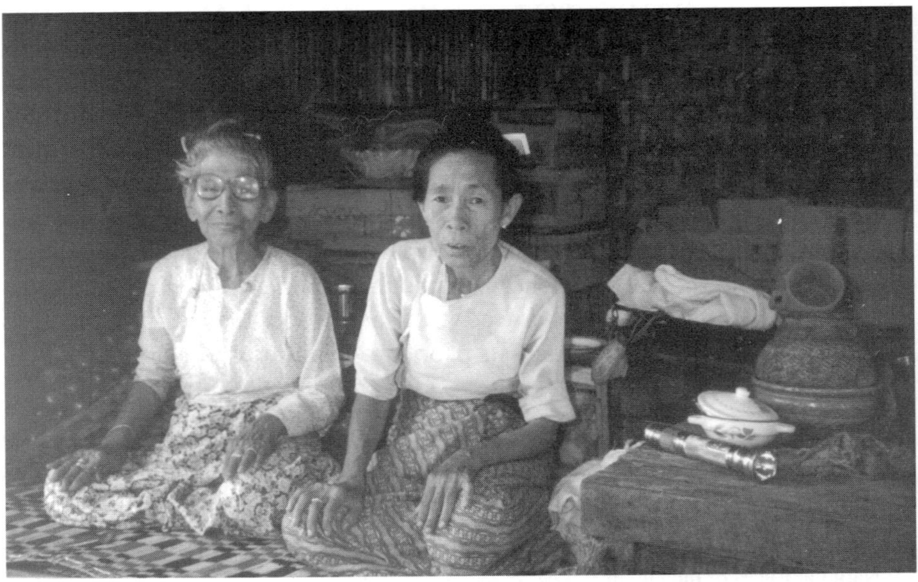

Inside the Home

British Grandeur

British Comfort

Traditional Couple

Boys and Men Wear *Longyi*

Historical Overview

The Pyu – a people who used the Sanskrit script and observed Buddhism with some elements of Hinduism – are the earliest people known to have flourished in this region. They were defeated by ancestors of today's Burmese. Based in Pagan, King Anawrahta – founder of the Pagan dynasty – ruled from 1044 to 1077, and it is under his rule that Buddhism spread across his unified kingdom.

Pagan

In 1287, the Mongols – under the leadership of Kublai Khan – arrived from China and destroyed the empire of the Pagan dynasty. The region was then ruled in the form of small states, dominated by three ethnic groups – the Burmese,[15] the Mons,[16] and the Shans.[17] This was so until the 16th century, when King Bayinnaung created a second Burmese empire. In 1752, King Alaungpaya established a third empire, which lasted until the arrival of the British. Traders from England arrived in the 17th century. Border troubles between Burma and British India led to war, which lasted from 1824 to 1826, when the British annexed the Burmese provinces of Arakan and Tenasserim. Following a second war, between 1852 and 1854, the British occupied Lower Burma. During a third war, in 1885, the British took over Upper Burma, making the former Kingdom of Burma a province of British India. While the Tibeto-Burmans and the Thai-Shans became British subjects, the Red Karens retained their special citizenship, in accordance with a treaty signed in 1875 (Christian, 1943). During the 1890s, Burma became an important exporter of oil.

Burmese Monks

---

[15] The Burmese were in control of Upper Burma since the 9th century.
[16] The Mons had come from the east, taking control of Lower Burma.
[17] Hagen (1962) traced the origins of the Shans to the Yunnan plateau of south-eastern China.

The Young Men's Buddhist Association, established in 1902, promoted the independence of Burma, as did Buddhist monks during the 1920s. A peasant uprising, in 1931, also put pressure on the British. On April 1, 1937, Burma became a colony separate from India. Outram & Fane (1940) give an account of pre-war Burma. Swerdlow (1995) noted that Burma's annual rice exports exceeded 3 million tons, making the nation the world's leading exporter of rice. Christian elaborated, "Burma is one of the few countries in Asia that have a considerable exportable surplus of food (1942, p. 3)." In December 1941, the Japanese invaded Burma, and by early 1942, the British withdrew. Burma had been an important base for Allied help to China, and the Japanese occupation of Burma was tragic for China, as the Burma Road was closed (Christian, 1943).

In 1943, Japan declared the independence of Burma, and set up a pro-Japanese government. After a long struggle (Arnold, 1944), Allied troops defeated the Japanese in 1945. On October 17, 1947, a treaty signed in London established the independence of Burma, and the Union of Burma came into existence on January 4, 1948 – a day of good omens, according to Burmese astrologers. The Karens[18] – a largely Baptist Christian minority with Chinese origins – waged war against the new government. Civil war lasted until 1952.

On March 2, 1962, General Ne Win pushed aside Burma's only elected prime minister and introduced military rule, along with the "Burmese way to Socialism," a blend of Buddhism, central planning and isolationism. Also in 1962, English was banned as "representative of a degenerate and decadent culture (Maung, 1991, p. 222)." The Burma Socialist Program Party became the only legal political party. Banks were nationalised, along with the transportation system, wholesale trade, retail trade, and foreign-owned companies. This led to a progressive breakdown of the formal firm-type sector of the economy. The state-controlled planned sector engulfed the national economy. The lack of foreign capital and technology resulted in limited growth during the 1960s.

In 1974, the Union of Burma became a single-party socialist state, and the one-party constitution renamed the country the Socialist Republic of the Union of Burma. The state confirmed its commitment to self-reliance, state control, isolation, and strict neutrality. In 1979, Burma isolated itself even from the movement of nations professing non-alignment. Steinberg (1982) gave a comprehensive socio-political account of the era.

[18] For a detailed discussion of the Karens, as researched by anthropologist Peter Kunstadter, see Kunstadter (1972).

From 1986 to 1988, declining trade contributed to a severe economic recession. Per capita income tumbled while inflation escalated. The author observed that between 1986 and 1987, military personnel quadrupled the price of petrol that they were selling in the parallel economy.

By 1987, Burma – formerly among the most prosperous economies of Asia – was designated by the United Nations as one of the least-developed nations in the world. Per capita income – which had been $670 in 1960 – fell to $190 in 1987, according to World Bank figures. In September 1987, people aspiring class mobility lost faith in their currency when the government invalidated all notes of 25, 35 and 75 kyat.[19] Falling living standards, coupled with political frustration, contributed to widespread riots during the first half of 1988. Although 77-year old Ne Win, Chairman of the Burma Socialist Program Party, resigned in July 1988, his party remained in power (*Herald Tribune*, 1988). In August 1988, tens of thousands of demonstrators toppled the dictatorship. In the absence of an entrepreneurial class, a new leadership arose from the military.

Following a coup d'état, on September 18, 1988, the State Law and Order Restoration Council took control of the country, and adopted a market-oriented policy, officially discarding central planning. On June 19, 1989 the government changed the English version of the country's name to the Union of Myanmar. During the early 1990s, about 6% of the GDP was accounted for by the manufacturing sector. Exports were encouraged, to obtain foreign exchange, but this resulted in local shortages, pushing up domestic prices. The mid-1990s encountered rampant inflation. A *pyi* – the local unit of weight equal to 2 kilograms – of rice that cost 11.72 kyat in 1991, was worth 41.11 kyat in 1993, and 60 kyat in 1996.[20]

In April 1997, the United States enacted economic sanctions against new investments in Myanmar. A few weeks later, on July 23, 1997, the nation joined the Association of South East Asian Nations (ASEAN). As Asia entered its financial crisis, the market rate of Myanmar's currency fell by nearly 100% in July 1997, and the dollar traded at more than 300 kyat. Meanwhile, the official exchange rate stood at about 6 kyat per dollar. In March 1998, new import restrictions were imposed. In 1999, approvals for foreign direct investment fell by 96%. Annual inflation during the 1990s averaged 27.13%.

---

[19] The unit of currency, spelled kyat, is pronounced "chat."

[20] In 1996, $1 was worth 5.75 kyat at the official exchange rate, or 125 kyat in the black market.

Toward Centrally Planned Transition

Colonial policy, in Burma, is said to have obstructed the development of local entrepreneurship. "Burmese enterprise was discouraged...the manufacture of mechanical lighters was made illegal, lest it should prejudice the revenue from a tax newly imposed on matches (Furnivall, 1956, p. 166-167)." The colonial government gave preferential treatment to those it favoured, including the British India Steamship Navigation Company, the British Overseas Airways Corporation (BOAC), the Burma Match Co. Ltd., and the Bombay Burmah Trading Corporation, Ltd. (Furnivall, 1957; Macaulay, 1934). Kyi (1970) also examined the impact of Westerners on the evolution of an indigenous entrepreneurial class.

Indigenous Entrepreneur

Despite colonial policy, there appears to have been an indigenous entrepreneurial spirit. "Contrary to the widely accepted assumption that European and Indian immigrants monopolized entrepreneurial roles in Lower Burma, the Burmese displayed considerable entrepreneurship in J. A. Schumpeterian's sense in that they put into effect new combinations in the means of production and credit (Adas, 1974, p. 210)." In particular, Burmese entrepreneurs were successful in the motion picture industry (Christian, 1943; Sein, 1950), printing (Swe, 1972), and rice milling (Hla, 1975). Burmese financiers were prosperous moneylenders (Wai, 1955), as were the Chettyars, who traced their origins to southern India. As well, Christian (1942) and Steinberg (1982) both pointed out that the Burmese were successful middlemen in the rice business. This is noteworthy, because elsewhere in the region it was the Chinese who dominated the middleman occupations.

Rice Business

Post-independence Burma – committed to self-reliance, state control, and isolation – adopted an economic policy that conformed to the import substitution model. In line with this model, the means to national economic development was to produce locally, thereby reducing the need for imports. This approach necessitated the existence of high tariffs, quotas and other regulatory mechanisms to discourage importation of goods. Rather than encourage specialisation in sectors of competitive advantage, this resulted in over-diversification. The model allowed the creation of a highly protected, low quality, high cost, inefficient industrial sector. Furthermore, exports were biased against, due to a typically overvalued official currency exchange rate.

As Ne Win resigned, in 1988, the Burmese Socialist Program Party was discussing proposals to reform the economy such that the "state would retain monopoly in commodities, broadcasting, the oil industry and gems and jade mining, but private enterprise would be allowed in transport, all kinds of industry and services, fishing, publishing and trade (*Herald Tribune*, 1988, p. 2)." It was the State Law and Order Restoration Council that initiated a market-oriented system, in 1988. Private enterprise was resumed that year. The Foreign Investment Law, enacted in November 1988, allowed foreign investment manufacturing, tourism and transportation. Other important reforms included the State-owned Enterprises Law of 1989.

The 1965 Law for the Establishment of a Socialist Economic System was revoked in March 1989. In 1990, further liberalisation measures were introduced to promote a market economy. As well, the Commercial Tax Law was introduced in 1990. The agricultural sector was stimulated by the reduction of controls and regulations relating to the cultivation, milling, storage, transport, and marketing of agricultural products. Records at the Planning Department of the Ministry of National Planning and Economic Development report that the economy grew by 3.7% and 2.8%, during the 1989-90 and 1990-91 periods respectively.

The Short-term Plan (1992–1996) had as objective to pursue an export drive and to achieve an average annual growth rate of 5.1%. Efforts were made to increase savings and to channel these savings into productive investments. According to San Khup, Director-General of the Planning Department of the Ministry of National Planning and Economic Development, actual growth rates – of 9.7% the first year, 5.9% the second year, and 6.8% the third year – were higher than projected (Khup, 1996). It may be argued, however, that these figures are distorted by the fact that inflation is underestimated.

During 1994 and 1995, government emphasis was focused on foreign investment and trade. Certain price controls were phased out, as were subsidies. Taxes were streamlined, banking was restructured, and international business was facilitated.

The Foreign Investment Law welcomed capital into Myanmar. Foreign-owned enterprises were permitted, as well as joint ventures with local enterprises. The law provided income tax holidays and further tax relief for exporters. It guaranteed repatriation of profits and stated that investments would not be nationalised. Private traders arrived from Singapore, South Korea and Thailand. Yet, a difficulty that remained was the inability of the state to mobilise domestic capital for investment. Another problem was the exchange constraint; although this encouraged barter, it limited trade.

In April 1995, the Rangoon City Development Committee made international news, by announcing the ban of the sale of betel in the nation's capital. This chewable nut, a mild stimulant that grows on areca palm trees, had been integral to the social and medicinal history in this country. Street-corner betel stands had provided employment for hundreds of people in Rangoon.

In March 1998, new import restrictions were imposed, and this led to a boom in the parallel economy. Myanmar's economic objectives of the 1990s included the development of agriculture as the base sector of the economy and the evolution of the market-oriented economic system. Yet, it was clearly specified that the initiative to shape the national economy must be kept in the "hands of the State and the national peoples."

At the Bazaar in Pyin Do Lwing, Formerly Maymyo

The Bazaar

Christian described bazaars in Burma, "Rubies, spinels, garnets…are bought and sold in open market…next to a bazaar selling potatoes, garlic, and Chinese radishes (1943, pp. 501–502)."

Typical of the bazaar, shops are clustered according to products or services sold. Jewellery shops stand adjacent to other jewellery shops. Each offers jade and fresh water pearls along with gold earrings and rings sold by weight. Likewise, fruit-sellers are clustered with other fruit-sellers. There are bananas and coconuts in abundance. Some vendors sell only tomatoes, while others specialise only in onions. Garlic is sold alongside ginger. Straw baskets contain different varieties of dry seeds. Rice comes in many varieties and different grades; some is damaged from poorly maintained husking machines. Cleaning, sorting and packing rice is for some women a social activity. Citrus fruits are not far from the tea specialist. A man is bargaining with a pharmacist as to the fair price of antibiotics. Ampicillin and tetracycline are available, as well as penicillin solution for injection; none of these require a prescription.

Selling Lotus Fruit

Food is plentiful. Palmyra palms are sold along with fresh apples, carrots, cauliflower, coffee beans, cucumbers, grapes, jackfruit, lotus fruit, mangoes, mangosteins, papayas, peanuts, plums, strawberries, sugarcane, and sunflower seeds. There are also dried fruits, and baked goods. As well, one finds pineapple jam, and mango jam, along with glass bottles of beverages. Lemon-barley is a favourite, as is green-coloured cream soda. Beasts of burden – cattle and water-buffalo – watch as chickens, ducks, goats, hogs, quail and sheep are sold alive. Lined up in a row, children are selling dry fish carefully laid out on the pavement. Lizards and stray dogs – friendly toward people, but at times territorial amongst themselves – scrounge for scraps. As the animals sense the oncoming storm, merchants act fast to pack up. Shoppers put their groceries on their head and make their way carefully to the waiting oxcart. The wind can be violent and bicycles can be sent flying. After the storm, the umbrella repairmen will have more business.

Strong Head

Oxcart

Stage-coach

As is the case with traditional merchandise, modern items are also clustered. Perhaps most notable are the moneychangers, who all use the same exchange rates. They buy and sell currency as if it were a commodity. Nobody has any competitive advantage. There is no product differentiation. The author observed several moneychangers clustered together: Ein Phyusi; 704; Bonton; Win; Yuzana; Crown; Forex; Gonshein Myint; User Friendly Stag; and May Flower – side by side. Although it is seemingly not possible to differentiate among them, each has his own loyal clientele. The focus here is not merely on transactions, but on the relationship between buyers and sellers.

Clustered Dealers in the Informal Sector

Personal Relationships Are Paramount

The Burmese Trishaw Holds Passengers Back-to-Back

Handicrafts

In 1857, King Mindon moved his capital to Mandalay and its Zegyo Bazaar became a principal distribution centre for beans, betel nuts, citrus fruit, cotton, nuts, onions, rice, tobacco, and wheat. The town prospered and artisans made it the handicraft centre of Myanmar.

Wooden Puppets Are Made in Mandalay

Among the crafts developed in Mandalay is *shwe saing* – the making of gold leaf. This occupation is concentrated in the Myatpa Yat Quarter, and found nowhere else in Myanmar (May, 1996). The skills are propagated from father to son, and non-family members are kept away from this occupation. The making of gold leaf is thus restricted to just a few families, propagated from one generation to the other. Rather than compete with one another, about forty entrepreneurs join forces to produce and sell gold leaf in a symbiotic manner. The co-ordination of activities, among the firms, facilitates an otherwise complicated process of production and marketing.

The first stage involves melting gold blocks; five households are involved in this initial phase of production, each with three or four skilled employees. Thirty households – each with about two-dozen workers are involved in the later stages of production. The molten gold is poured into a conical vessel, producing finger-shaped bullions. Rollers flatten these bullions, which are subsequently beaten by skilled craftsmen resulting in thin sheets of gold, referred to as *shwe-let* – literally "golden fingers." Each sheet – about 2½ cm wide and over one metre long – takes about 4½ hours to flatten. These are then sent to another household for further processing. The gold is cut further, into pieces about 25mm by 5mm. These are placed between *shwekhat setku* – sheets of bamboo paper made for the purpose of producing gold leaf. The artisans then wrap layers of this in deer-hide, and the gold is beaten again, to become film-thin. Once this is done, the strips of gold are joined with one another, to form quadrangles – each about 25 square centimetres. The final product is stored in *shwehlaung setku*, a paper made in Amarapura, from straw. Much of the gold leaf is used by religious people, to gild pagodas and images of Buddha. As is the production of gold leaf, stone carving is also a hereditary occupation. This activity is clustered in the Kyauk Sit Tan Yat Quarter of Mandalay. As is the case with the production of gold leaf, this industry is also intertwined with Buddhism and much of the carving – in marble and in wood[21] – involves making images of Buddha.

Images of Buddha are also cast in bronze, and bronze casting is important in Myanmar. In Amarapura, is an area called the Bronze Village, where bronze casting works are clustered. Bronze and copperware are often produced by trained silversmiths who are lacking silver, a scarce metal in Myanmar. Silversmiths are clustered in Ywataung, not far from Sagaing.

---

[21] Puppets are also carved in wood.

Mandalay

Along the Ayeyarwady River at Mandalay

Silver and gold are used to make embroidery in Myanmar. In contrast to gold leaf, which is produced for consumption by local people, embroidery – containing gold and silver thread – is used to make tapestries, fabrics and cushion covers, for sale to tourists visiting Myanmar.

Another craft that thrives in Mandalay is silk weaving. Silk – mostly imported from the PRC – is spun into thread, which is then boiled in dyes to get coloured. It takes about ten days to weave a metre of cloth. Firms active in this industry are clustered east of the royal palace, in a neighbourhood of Mandalay called Nan Shay.

Carving in Mandalay

Embroidery

Informal Business Activity: Fast Food for Train Passengers

Covert Economic Activity

According to sources at the United Naticns Office for Drug Control and Crime Prevention (UNODCCP), Myanmar is the source of 23.4% of the 4,653 tonnes of opium produced around the world each year. It is from these opium poppies that heroin is made. The wholesale price per kilogram of heroin, in New York, is $80,000, according to sources at *The Economist*, and at 40% purity, the retail price is $290,000. Production is inexpensive, and the profit margin is immense. Its profitability makes it hard to control, because the hefty contribution margins make it easy to bribe poorly paid law enforcement officials.

Pluralism

According to the Ministry of Information, there are 135 national groups in Myanmar, with the Bamar being the majority. The different ethnicities are classified according to distinct ethno-linguistic groups. While the majority of the people are Tibeto-Burman, the country is also home to Mon-Khmers, and Thai-Shans, as well as Karens. The Tibeto-Burmans speak a Sino-Tibetan language, as do the Karens. In contrast, the language of the Mon-Khmers belongs to the Austro-Asiatic family, while the Thai-Shans comprise one of the Thai groups.

In addition, there are Chinese and Indian minorities in Myanmar. It was the British who encouraged the immigration of Indians to Burma. Some became moneylenders, and others were given government positions. By the 1930s, the immigrant Indian community in British Burma numbered 600,000, amounting to less than 10% of the population; yet these people held over half the government jobs.

Ethnic consciousness is notable. Dunung noted, "Burmese having Westernized attitudes are looked down upon. The Burmese also consider themselves to be superior to Indians, Chinese, and Eurasians (1995, p. 528)."

Elephant Labour

Myanmar is home to as many as 6,000 wild elephants, and up to 7,000 domesticated ones (Gray, 2001). The Myanmar Timber Enterprise – which has a monopoly on teak extraction – employs 2,700 elephants. These are very useful in areas where vehicles cannot penetrate without causing major ecological damage.

A 30-year cycle is used to harvest teak, and only trees of a designated girth are felled within a given area. Given this selection process, trained elephants do a neater job than would be possible if using machinery.

An elephant starts school at the age of three. Students can learn one command every two days. Classes alternate with recess for bathing and resting. Once on the job, a bull elephant can work five to eight hours per day, depending on the temperature. The animals use their husks, forehead and trunks to push literally tons of logs at a time. As well, these intelligent workers are made to drag trees estimated to be 150 years old. Retirement age is 55.

Elephant Travelling to the Job, by Truck

Water-buffalo Also Haul Logs

Internal Economic Activity

Life in Yangon

Toward the Future

Since 1988, steps have been taken to liberalise the agricultural sector, legalise border trade, and encourage foreign investment. Firms in Myanmar are now involved in agricultural processing, light manufacturing, production of footwear, pharmaceuticals and textiles. Their export markets include China, India, Singapore and Thailand.

Yet, as reported by Speece and Sann (1998), managers complain about labour shortages and the lack of loyalty among employees. Not surprisingly, most large companies have a minimal presence in Yangon, and much investment is small scale.

Many local enterprises operate at the subsistence level. A self-employed individual sells air for vehicle tyres. Another fixes umbrellas on a street corner, and on the same broken sidewalk, a barber has set up an impromptu stall. A tourist is being shaven for the price of a coconut. Nearby, a micro-scale merchant sells flashlights and lighters. Another vendor is weighing nails. A woman walks by on her way to the station; she has prepared food to sell to passengers on the train.

At a pagoda, a woman sells sparrows to worshippers who will free them for good luck. As soon as they are let loose, she catches them again, as her inventory is fixed. Meanwhile, the parallel economy is thriving with consumer goods and pirated intellectual property. Near an outlet called Mister Donut – the sign of which uses the font of Dunkin' Donuts – a merchant sells photocopies of the book *Burmese Days* (Orwell, 1930). Not far, a Ronald McDonald look-alike promotes Mac Burger, a fast-food outlet serving MacHam, MacChicken, and MacBurgers. A client takes out his "Solem" cigarettes from a pack that looks like that of Salem. An ambulant vendor heads for a bus on which she hopes to sell quail eggs; she is careful not to step on the mangoes being laid on the sidewalk by a micro-scale vendor. What is consistent, throughout all this, is the absence of tension.

While the government has focused on improving agricultural output, increasing the production of land, Myanmar remains among the less developed countries. Yet, the extent of absolute poverty is limited due to the abundance of inexpensive food and other essentials.

Chapter 11

An Overview of Indochina

Introduction

The following three chapters are about Cambodia, Laos and Vietnam[22] – formerly known collectively as Indochine française – French Indochina. There are some important differences, as well as similarities, among the constituents of this region. As explained by Moore & Williams, China and India were the cultural parents of Indochina, making it a "mosaic of Indian and Chinese cultures (1951, p. 462)." Hindu Brahmans sparked the native genius of the early Khmers, while Chinese officials lent their culture to the Annamese. Under colonial rule, the whole region used the same currency, the Indochinese piastre – issued by the Bank of Indochina.

It would be naïve, however, to overlook significant differences within this region. Traditionally, in Cambodia, the ownership of land was strongly individual in nature. In contrast, Vietnam had a tradition of communally owned land. Long observed:

> Unlike Vietnam, which took China as its teacher, Cambodia and Laos received much of their art, religion, and language from India. Both countries resemble Thailand more than they do their Indochinese neighbor...Nor has Cambodian writing been changed to Roman letters like Vietnam's; it uses graceful, rounded lines of Sanscrit (1952, p. 312).

---

[22] Literally, Vietnam means "people of the south."

Graceful, Rounded Lines in Cambodia

The Lao Language Has its Own Script

Roman Letters in Vietnam

Indeed, in contrast to the adjacent area – corresponding to southern Vietnam – which had its status changed to that of a colony and its alphabet changed to the Roman one, Cambodia had relatively little Occidental influence. Cambodia retained its monarchy and its script.

Royal Palace in Phnom Penh

Zasloff (1988) examined post-war Indochina. Vietnam became independent on September 2, 1945, Laos on July 19, 1949, and Cambodia on November 9, 1953.

In the 21st century, there are still important differences to be found when comparing the Kingdom of Cambodia, the Lao People's Democratic Republic, and the Socialist Republic of Vietnam. Each of these countries has implemented its own model of transition, reflecting different paths to the future.

Vietnam Solidarity Monument in Cambodia

Different Models in Former Indochine Française

In Cambodia, a true veritable transition has taken place. This involves complete "restructuring" under the constitutional monarchy. The Kingdom of Cambodia has already privatised most state-owned enterprises, and import licenses have been abolished for most items. O'Driscoll, Holmes & Kirkpatrick (2001), described Cambodia as being freer[23] than any other transition economy of Asia. They showed that the government in Cambodia is less of a fiscal burden than is the case in the other countries of Indochina. Nevertheless, as observed by Prasso, "compared to neighboring Laos and Vietnam, which have also struggled to recover from an era of war and bloodshed, Cambodia remains far behind in its ability to establish the most basic building blocks of economic recovery...Domestic savings rates are stunningly low compared to other Asian countries (2001, p. 2)."

[23] Beach & O'Driscoll defined economic freedom, as "the absence of government coercion or constraint on the production, distribution, or consumption of goods and services beyond the extent necessary for citizens to protect and maintain liberty itself (2001, pp. 43-44)."

O'Driscoll, Holmes & Kirkpatrick (2001) show that Cambodia has less black market activity than any other transitional economy in Asia. They also reveal that the economies of Laos and Vietnam each have more government intervention than is the case in Cambodia. While Cambodia has fewer restrictions on foreign direct investment, than any other country described in this book, Laos has the most.

In contrast to the situation in Cambodia, the leadership of Laos is still recognised as being communist, and even transition is centrally planned. Weighted by bureaucracy, nothing changes fast in Laos.

In Vietnam, the Communist Party has taken steps, to "renovate" the economy with some transitional measures; yet, there has been no attempt to "restructure." The state has been slow to implement structural reforms, and this has been a concern for investors. Foreign direct investment fell from a high of $8.3 billion in 1996, to a total of $1.6 billion in 1999.

Heterogeneity

Cambodian law includes royal decrees, as well as legislation influenced by the United Nations Transitional Assistance in Cambodia (UNTAC). In contrast, the legal system of Laos is largely based on traditional customs and Soviet practice. The law in Vietnam blends communist theory with the French Civil Code.

Cambodia's merchant marine is seemingly large, as the government offers a flag of convenience to owners of foreign vessels. In contrast, Laos has one ship registered over 1,000 GRT. Vietnam, in 2002, had about 150 Vietnamese-owned ships registered.

In Cambodia, literacy among males is 48%; among females, it is 22%. In Laos, the figure is 70% for males and 44% for females. Vietnam is relatively more educated, with a literacy rate of 97% among males and 91% among females. Not surprisingly, the level of poverty varies too. In Cambodia, 36% of the people live below the poverty line, while the statistics for Laos and Vietnam are 46% and 37% respectively. This is summarised in Table 15.1, in Chapter 15.

In addition to differences among the countries of Indochina, there is also heterogeneity within each country. Most noticeable is the ethnic Chinese minority, very active in enterprise. Over half a century ago, Moore stated, "keen traders, Indochina's 500,000 Chinese dominate many businesses (1950, p. 508)." Central planning has since come and gone, and the Chinese are still involved in entrepreneurial activity.

Cyclos in Cambodia

In Vietnam, the Driver Sits Behind the Passenger(s)

In Laos, the Passenger(s) Sit Behind the Driver

Train Station in Phnom Penh

Unlike Laos, Vietnam Has a Sophisticated Rail System

Chapter 12

The Kingdom of Cambodia[24]

Introduction

Cambodia covers 181,035 square kilometres. It lies on the Gulf of Siam, bordering Laos, Thailand, and Vietnam. Cambodia is among the least developed countries in the world. Traditionally, the people of Cambodia – known as Khmers – were not inclined to become entrepreneurs, as enterprise was not seen as contributing to Buddhist society. A merchant class developed after independence, but during the 1970s, the Khmer Rouge – officially the Communist Party of Kampuchea – extinguished private enterprise in this country. More recently, the ownership of property was legalised, and the nation embarked on a comprehensive programme of transition to a market economy.

According to O'Driscoll, Holmes & Kirkpatrick (2001), Cambodia ranks higher in economic freedom than any of the following: China, Kazakhstan, the Kyrgyz Republic, Laos, Myanmar, Tajikistan, Turkmenistan, Uzbekistan and Vietnam. The same study also calculated average tariff rates and found Cambodia to have the lowest of all these countries. As well, that study noted less government intervention in Cambodia than in any other country on the above list; comparing the fiscal burden of governments, Cambodia scored lowest.

---

[24] This chapter includes information obtained from the Ministry of Commerce and the Ministry of Finance and Economy. The chapter also includes material that first appeared in Dana (1999c).

Traditional Khmer Home

French Colonial Mansion in Phnom Penh

Historical Overview[25]

About 2,000 years ago, much of that which is Cambodia today belonged to the Kingdom of Funan – a prosperous nation frequented by merchants travelling between India and China (for a detailed account, see Chandler, 1999). At the time, Mahayana Buddhism flourished alongside the religion and culture of the Hindus.

During the 6th century, the Kambu'jas – literally "those born from Kambu" – whose relationship with Funan is obscure, established a kingdom in the interior of present-day Cambodia. Chinese travellers knew this kingdom as Chenla. Its people associated themselves with Kambu, a figure of Indian mythology and the nation became known as Kampuchea, a derivative of Kambu'ja.[26]

In 802, a Khmer prince declared himself a universal monarch, and founded a dynasty that lasted until Angkor was abandoned in the 16th century (Chandler, 1999). During the Angkorian period – 9th to 15th century – the civilisation stretched into parts of today's Laos, Thailand and Vietnam. Indravarman, who ruled from 889 to 910, built an elaborate irrigation system. Jayavarman VII, who ruled from 1181 to 1201, built a new capital – Angkor Thom.[27] During the 13th century, the nation gave up Hinduism, in favour of Buddhism. The official language of the Hindu priests was also abandoned.

Over the next two centuries, several tributary states – in what is now Thailand – declared their independence and invaded Cambodian territory (Chandler, 1999). The capital city shifted away from Angkor to the region of Phnom Penh.

In 1860, King Norodom ascended to the throne. Fearful of nearby Siam, in 1863, he asked France to provide protection, and he subsequently signed a treaty with Napoleon III, making Cambodia a French protectorate. In 1904, King Norodom was succeeded by King Sisowath, who reigned until 1927.

---

[25] This section benefited greatly from the personal assistance of Professor David P. Chandler, PhD, Professor Emeritus of History, at Monash University, and Visiting Professor at Cornell University. Professor Chandler is recognised as the leading authority on Cambodian history.

[26] The Portuguese later modified *Kambu'ja* to *Camboxa*. The French adapted it to *Cambodge*. When anglicised, the name became "Cambodia."

[27] *Angkor* means "the city," in Khmer, while *Thom* means "large."

King Sisowath was succeeded by King Monivong, who died during the early phases of Japanese occupation. Prince Norodom Sihanouk – born on October 31, 1922 – was crowned king, in 1941.

After the Japanese surrender, in 1945, France made Cambodia an autonomous state within the French Union – but retained control over its diplomatic, financial and military affairs. In 1953, the *Royaume du Cambodge* – Kingdom of Cambodia – declared its independence, which was recognised in May 1954, by the Geneva Conference. Self-sufficient in food, the new kingdom enjoyed economic growth and became a net exporter of rice (Ear, 1995; Prud'Homme, 1969; Steinberg, 1959). In 1955, Sihanouk abdicated, but remained for 15 years the leading political figure in Cambodia.

Although Cambodian-American relations deteriorated after Cambodia rejected the American economic and military aid programme, in November 1963, Cambodia remained officially "neutral." Radio Phnom Penh subsequently stepped up anti-American propaganda (Abercrombie, 1964). For a lucid account of Sihanouk's policy of neutrality, see Hamel (1993).

In 1969, the United States began bombing suspected communist camps in Cambodia. The population of Phnom Penh was slightly more than half a million at the time, and Cambodia was agriculturally prosperous.

In March 1970, General Lon Nol staged a successful coup d'état. Sihanouk was deposed and he moved to Beijing. The country was renamed the Khmer Republic. Chantrabot (1993) gives a detailed account of that era.

On April 30, 1970, American and South Vietnamese troops attacked Cambodia. During the following years, hundreds of thousands of people were killed, and many fled from rural areas to Phnom Penh. By 1975, the capital city was home to 2 million people.

On April 17, 1975 the Khmer Rouge, a group whose goal was to transform the nation into an agrarian peasant-dominated Maoist co-operative, overran the republic. Under the leadership of Saloth Sar – who used the pseudonym Pol Pot (see Chandler, Kiernan & Boua, 1988) – the Khmer Rouge transformed the nation (Jackson, 1989), as the new regime confiscated private property, banned currency and changed the name of the country to Democratic Kampuchea. Millions of city dwellers were ordered to evacuate their homes and to migrate to the countryside; they were told that these measures were being taken in order to avoid bombings by the United States.

The Khmer Rouge Evacuated Urban Areas

Scars of the War

Cambodians interviewed by the author explained that the Khmer Rouge communists slit the throat of their prisoners, in order to save ammunition. Women's nipples were removed with pliers, and victims were left to scorpions.

Border clashes between the Khmer Rouge and the Vietnamese led to an invasion by Vietnam on December 25, 1978. On January 7, 1979, Vietnam toppled the Khmer Rouge regime. Again, the name of the country was changed, this time to the People's Republic of Kampuchea. Currency was re-introduced and people were again permitted to reside in towns.

Quarantine Service of the People's Republic of Kampuchea

Since the Khmer Rouge had killed the urban landlords, Phnom Penh was desolate at the time. Nevertheless, it was quickly populated as anybody could claim whatever property they wished, on a first come, first serve basis. People returned to Phnom Penh, but infrastructure was lacking.

Rice fields were abandoned as masses flocked to urban areas. Famine followed. During the 1980s, the USSR and Vietnam contributed food to Kampuchea. That same decade, the government devalued the Kampuchean currency. In 1984, $1 bought 7 riels; by the end of the decade, $1 was worth 380 riels.

In September 1989, Vietnam withdrew its troops, and the State of Cambodia was established. However, the Khmer Rouge kept a hold of gem-rich lands neighbouring Thailand. Rumours spread that currency might be banned again and people dumped riels to buy gold; the currency tumbled.

In 1991, the United Nations brokered a peace accord known as the Paris Agreement, which was signed in October that same year. On October 31, the United Nations Security Council agreed to establish the United Nations Transitional Assistance in Cambodia (UNTAC).

UNTAC in Phnom Penh

African Peacekeepers in Cambodia

The Khmer Rouge continued fighting in the jungle, and in January 1992, riots broke out in Phnom Penh – with the return of nominal Khmer Rouge leader, Khieu Samphan. On February 28, 1992, the United Nations Security Council voted to send a force of 22,000 to Cambodia.

Although the signing of the peace accord strengthened the riel, the influx of United Nations peacekeeping troops contributed to higher prices and subsequent devaluation. Retailers and even government ministries began refusing local currency. Inflation, in 1992, was 177%, while per capita GNP was $150. Government employees were earning low wages and often pursued sideline opportunities.

The 1993 constitution re-established the monarchy. Norodom Sihanouk was crowned king, once again, on September 23.

In 1994, the literacy rate edged up to 35%. That year, the dollar traded at 2,400 riels, inflation fell to 26%, but fighting escalated. Brown & Zasloff (1999) provide a detailed account of peacekeeping efforts.

Although the Khmer Rouge continued to control its stronghold – in the northwest of the country – economic growth approached 5% in 1995; per capita GNP was $215, while inflation stood at 3.5%. The average value of $1 was 2,451 riels in 1995, 2,624 riels in 1996, and 2,938 riels in 1997.

King Norodom Sihanouk

During the first weekend of July 1997, a coup d'état by Hun Sen – holding the position of Second Prime Minister – overthrew Prince Norodom Ranariddh, the so-called Prime Minister of Cambodia at the time. The nation was scheduled to join ASEAN – along with Laos and Myanmar – on July 23, 1997. However, on July 10, the members of ASEAN decided to postpone admitting Cambodia into the association. The last Khmer Rouge fighter surrendered in December 1998. On April 30, 1999 Cambodia officially joined ASEAN. The other members were Brunei, Indonesia, Laos, Malaysia, Myanmar, the Philippines, Singapore, Thailand and Vietnam.

The Royal Cambodian Model

Under colonial rule, Cambodia's public policy was dictated by France, which was actually more interested in neighbouring Cochin-China. Taxes raised in Cambodia were used to develop Cochin-China. In Cambodia, the French left the economy to market forces. Large-scale rubber plantations – which belonged to Europeans – imported workers from elsewhere, in Indochina, justifying this by a claim that the local people were less productive. Since all of French Indochina used the same piastre, this facilitated such "imports."

French control ended with the independence of Cambodia. The *Banque National du Cambodge* (National Bank of Cambodia) introduced the local currency – the riel, which was pegged to 23.3905 mg. of gold. From 1953 to 1970, Sihanouk implemented a policy of national socialism. In the 1955 elections, the *Sangkum Reastr Niyum* (People's Socialist Community) won every seat in parliament. Although socialist policies were introduced, about 80% of the farmers remained the proprietors of the land that they farmed. Most of the farmers cultivated rice, and many raised animals. In urban areas, the ethnic Chinese controlled the economic sphere; 77% of these were Teochew Chinese from Guang Dong Province in China.

Sihanouk introduced his first economic plan in 1956. It was a two-year plan, the budget of which the United States paid 57%, China 23% and France 17%. Only 3% of the budget was raised by Cambodia. The two-year plan emphasised the role of capital, but ignored private investment and entrepreneurship. The plan was renewed for another two years, beginning in 1958. After that, the Planning Ministry began to set longer plans. The first of these covered 1960 to 1964.

In 1963, private banks were nationalised and austerity measures resulted in many tariffs being doubled. A new economic plan covered the period 1964 to 1968 and another from 1968 until 1972.

In April 1975, the Khmer Rouge implemented an unprecedented policy reform, with the goal of transforming the country into a peasant-dominated, agrarian collective. During the last two weeks of April, the Khmer Rouge evacuated Phnom Penh, forcing every urban resident out of the city and into slave-labour camps in the countryside.

The regime proclaimed 1975 as Year Zero (Ponchaud, 1977). Private property, in theory and in practice, ceased to exist. The Central Bank was blown up, other banks were closed, currency was abolished, postal service came to a halt, and the regime caused the death of over 1.5 million[28] Khmers – out of a national population of about 7 million – eliminating intellectuals, landlords, entrepreneurs and the business sector. Market activities were banned and all commercial transactions were outlawed. In addition to a change in clothing, individuals were allowed only two items of private property, namely a bowl and a spoon. The urban areas, including the capital city, Phnom Penh, were depopulated and would remain empty for over three years. All flights to and from Democratic Kampuchea were halted – except a fortnightly flight linking Phnom Penh with Beijing. The regime designed a four-year plan for the period 1977 to 1980. In 1977, people were required to participate in communal cooking and eating, as Pol Pot banned the private ownership of pots and pans.

When the Khmer Rouge regime was driven from power in 1979, trade resumed, but in the absence of a national currency; commerce relied on barter or foreign money. Only in March 1980 was currency reintroduced, and then only within the context of a socialist economy. Cigarettes, condensed milk, kerosene, rice, soap and sugar were subsidised by the state, but goods were rationed.

After the 1989 withdrawal of the Vietnamese from Cambodia, reforms enhanced the economic environment for entrepreneurs to operate in the private sector. The local authorities divided farmland and distributed it to those living on it. In the urban areas, the Law of Private Ownership allowed persons to claim title to property occupied, and even to sell it.

---

[28] Vietnamese sources claim the figure to be 3 million.

Return to Private Ownership

Budding Entrepreneur

In August 1991, the National Assembly passed the Law on the Management of Exchange, Precious Metals and Stones. This legislation contained 20 articles governing foreign exchange, precious metals and gems. In November, the Foreign Exchange Decree supplemented this law.

By 1992, three quarters of the economy was in private hands. Some entrepreneurs operated from their homes, while others lived in their shops. Smaller-scale vendors set up impromptu stalls. Petrol was sold on roadsides, in used soda-pop bottles.

Entrepreneurs were required to pay a signage tax and, therefore, many merchants – including prominent retailers in Phnom Penh – opted to have no sign.

Few Signs

A new constitution came into effect in 1993. Article 56 specified that the kingdom would have a market economy.

In an attempt to improve the inadequate infrastructure, in November 1993, the kingdom participated in the formation of Royal Air Cambodge, to supplement the operations of state-owned Kampuchea Airlines. Royal Air Cambodge was established as a joint venture, 60% owned by the government and 40% by a Malaysian affiliate of Malaysia Airlines (MAS). Linking Phnom Penh with Bangkok, Guangzhou, Ho Chi Minh City, Hong Kong, Kuala Lumpur and Singapore, Royal Air Cambodge became the largest airline in Cambodia. Yet, between 1994 and 2001, the firm lost over $30 million. In September 2001, flights to Ho Chi Minh City were suspended. On October 16, 2001 the airline stopped flying.

Kampuchea Airlines Tupolev TU-134A

In August 1994, liberal laws were implemented to encourage entrepreneurship in the kingdom. A problem, however, was that a very lenient implementation of public policy allowed latent corruption to persist. Also in 1994, special zoning laws were introduced, to limit development in the area of Siem Reap; only small-scale ventures were allowed near Angkor Wat, and large hotels were required to be four kilometres away. Nevertheless, entrepreneurs could bribe officials and obtain exemptions. In 1997, an article (Desjardins, 1997) in the French daily *Figaro* dubbed Cambodia the "Kingdom of Corruption." At the time this book went into press, there was still no anti-corruption law in Cambodia.

Pluralism

The Chinese always had a significant impact on the economy in Cambodia. Gold, silver and coins were introduced to Cambodia by the Chinese, and Chinese traders flourished at Angkor. Immigrants from China came to dominate the business realm.

Traditionally, the ethnic Khmers tended to be attracted to occupations related to agriculture, government service, and monastic life; commerce and industry were occupations that were left to ethnic minorities. Muslims controlled the cattle trade, commercial fisheries and the weaving industry. The ethnic Chinese engaged in international trade and in retailing.

Ethnic Khmers

Abercrombie reported on the situation during 1964:

> Nearly a third of Phnom Penh's population of 403,500 are Chinese. Together with Vietnamese importers, shopkeepers, and moneylenders, they dominate the city's – and the country's – commerce. In the shops strange goods caught my eye: dogmeat sausages, incense sticks, begging bowls, silver elephants, betel leaves, Chinese comic books, brass hongs, and bamboo flutes (1964, p. 518).

Notice that the above account does not mention ethnic Cambodian entrepreneurs. Elaborating on the ethnic Chinese, in Cambodia, Abercrombie explained, "Rarely holding citizenship, they maintain their own language, customs, and schools (1964, p. 526)."

Today, Cambodia is among the most homogeneous countries in Asia. There are, nevertheless, relatively small minority groups. The ethnic Chinese community is concentrated in entrepreneurial activities. The Vietnamese in Cambodia are often associated with fishing-related occupations. Although both of these minorities have suffered persecution in Cambodia – especially the Vietnamese – the Chinese are emerging as an important economic force, and Chinese aid is crucial to the well-being of the current regime. Cambodia is also home to some 200,000 Cham Muslims and members of the ethnic hill-tribe, Khmer Loeu.

The Nature of Entrepreneurship

Entrepreneurship in Cambodia developed rapidly after independence from France. In 1955, there were 650 small and medium-scale factories in Cambodia. By 1968, there were almost 4,000. Most entrepreneurs, however, were not industrialists. Many were speculators who tried to make fast money in alcohol, beef, gold, land, tobacco, salt and other commodities. These people also contributed to economic instability.

The economy collapsed during the era of the Khmer Republic, which lasted from 1970 to 1975. Manufacturing declined due to shortages of inputs. The republic financed its deficit by printing money and imposing a 60% Value Added Tax (VAT) on cigarettes.

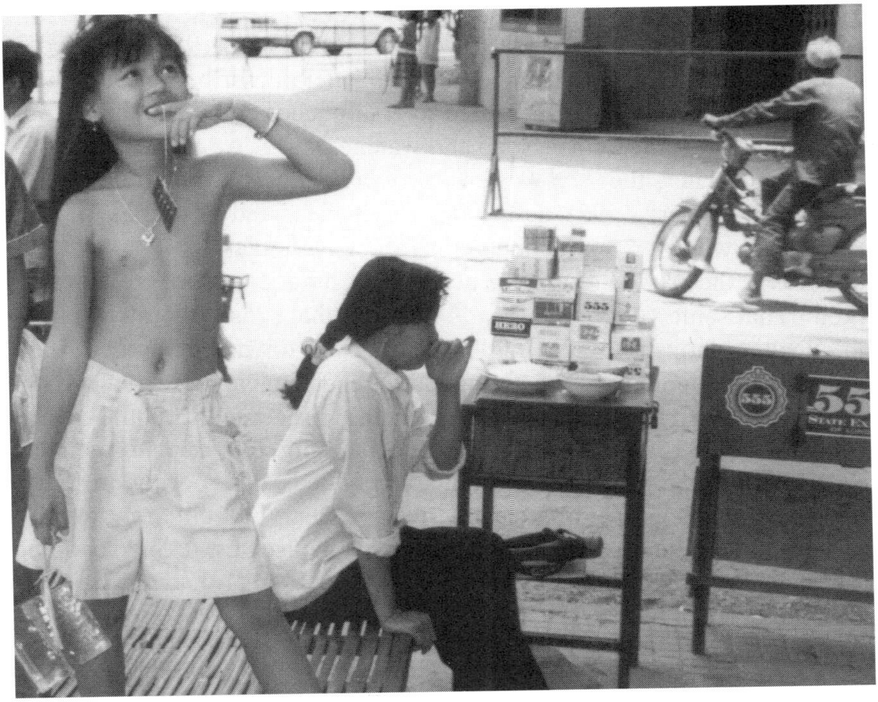

Cigarette Stand

Fast Money

In May 1982, a *National Geographic* essay (White, 1982) reported on covert economic activity in Cambodia. The article described entrepreneurs dealing in contraband goods smuggled to Kampuchea from Thailand. These included cigarettes, medicines, soap and watches.

A decade later, the *Far Eastern Economic Review* ran a cover story (Chandra & Tasker, 1992) describing the widespread activities of Thai entrepreneurs who were extracting gems from Cambodia.

In 1995, the *Washington Post* described Cambodia as a "pre-emerging market." That same year, *Paris Match* reported that prostitution was Phnom Penh's leading industry, with 10,000 prostitutes in town and another 10,000 elsewhere in the country. The prominent French magazine explained that virgins were rented for 2,500FF[29] before being discounted to 200FF. While opium was worth 150FF per gram, a bag containing 200 grams of cannabis cost 10FF. Marijuana is legal, and commonly added as flavouring, in soups.

Meanwhile, some entrepreneurs made 6-year old children work in factories. Other children became entrepreneurs themselves, taking advantage of arbitrage opportunities, selling food and drinks to long-distance travellers on boats, buses and trains.

Young Entrepreneurs

[29] The French franc was the currency of France until the introduction of the euro (€) on January 1, 2002. The euro was valued at 6.55957 FF.

Desjardins (1997) described how government employees in Cambodia were selling, for personal gain, rubber and wood that belonged to the state, in addition to medicines donated by foreign parties, and meant for free distribution.

On March 7, 1999, the *Japan Times* (1999) described poverty-driven Cambodian entrepreneurs dealing in rare animals and their products. In Phnom Penh, an entrepreneur offered an endangered sun bear for sale, for $600.

Cashew Nuts Are Common

Not So Common

Toward the Future

This chapter began by summarising findings of O'Driscoll, Holmes & Kirkpatrick (2001). Compared to China and to other transitional economies, Cambodia ranks relatively high in economic freedom. Average tariff rates are lower in Cambodia than in China, the Kyrgyz Republic, Kazakhstan, Tajikistan, Vietnam, Myanmar, Turkmenistan, Uzbekistan and Laos. There is less official government intervention in Cambodia than in any of these countries. Comparing the fiscal burden of governments, Cambodia scored lowest on the list. Yet, further investigation suggests that this may be an optimistic picture. Naranhkiri Tith, Professor of International Economics and Asian Studies, at John Hopkins University, explained to the author: "Tax burden is low because of the inability to tax and not because of the willingness not to burden the tax payers; tariff is low because room must be left for corruption…and the official government intervention is less but corruption—a form of unofficial government intervention – is pervasive."

Not surprisingly, although most enterprises in Cambodia are privately held, entrepreneurship in this kingdom is concentrated in grey or black market areas. Casinos and shady banks are said to launder money. Drugs are traded openly.

The historical experience of this nation makes immediate gains more attractive than waiting for an uncertain future. All Cambodians over 35 years old can remember the day money was made useless; the result is a predisposition to reject currency as a medium for savings, and for large purchases, Cambodians often use the damleung (1.2 ounces of gold) to buy property, and the chi (0.13 ounces of gold) for items such as televisions (Prasso, 2001).

Not surprisingly, entrepreneurship in Cambodia is largely short term in nature, due to high risks and uncertainty. People use a young tree for firewood, as they are impatient to wait seven years for it to produce rubber. A more long-term orientation would be desirable in the future. In Cambodia, however, institutions are weak and traditions are strong.

Chapter 13

The Lao People's Democratic Republic[30]

Introduction

The Lao People's Democratic Republic – also known as Laos – is one of Asia's most undeveloped nations, and among the five poorest in the world. Perazic wrote, "Once a great kingdom, 'the Land of the Million Elephants and the White Parasol' faces its struggle for survival (1960, p. 46)." The landlocked country, enclosing 236,800 square kilometres, is nestled between Cambodia, China, Myanmar, Thailand and Vietnam. Laos is sparsely populated. It is among the least urbanised nations in Asia.

The Hammer & Sickle Still Symbolise the Economic Model of Laos

---

[30] This chapter includes information obtained from the Ministry of External Economic Relations and the Ministry of Industry and Handicraft. The chapter also includes material that first appeared in Dana (1995b; 1999c).

Traditionally, business activities in Laos were not associated with high social status. Cultural values, stemming from religious beliefs, emphasised, instead, the elimination of desire. Commerce, on the other hand, was perceived as a means to satisfy desire. Social forces thus discouraged enterprise, and trade was usually the role of those with inferior social standing. When Lao men refrained from doing business, women often succeeded (Dana 1997e). Yet, the communist take-over further discouraged entrepreneurial spirit. As a result, Lao society is generally non-entrepreneurial. The Chinese community of 67,000 people – 1.3% of the population – is very active in the entrepreneurship sector. Large corporations in Thailand, each earn more than the value of all the goods and services produced in all of Laos.

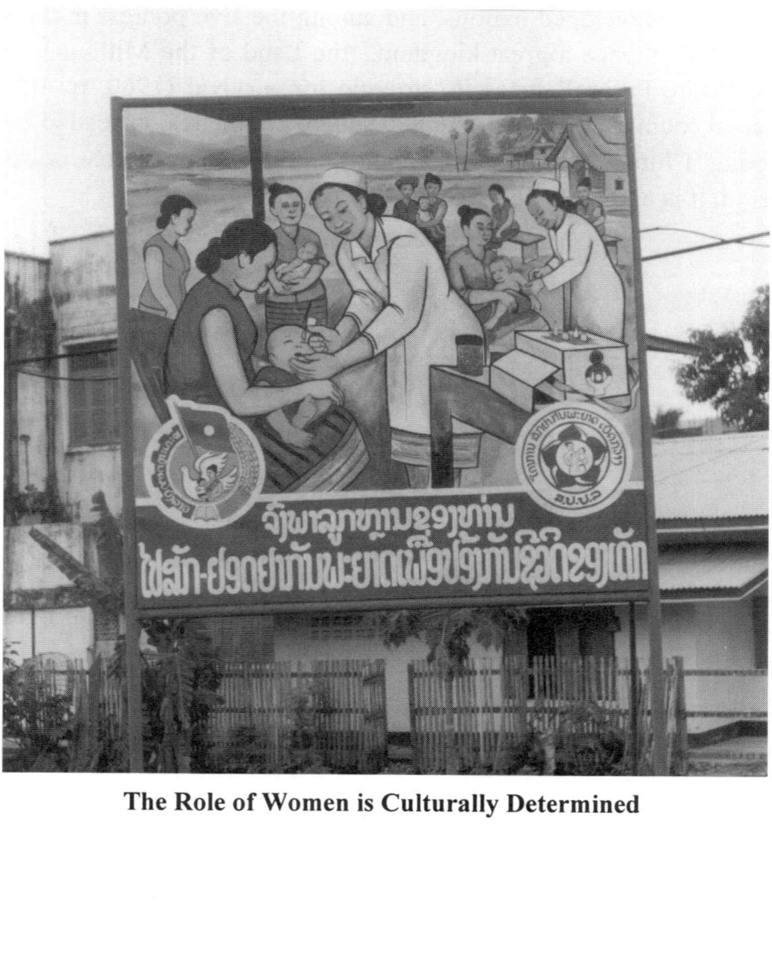

The Role of Women is Culturally Determined

Historical Overview

The Lao people are thought to have arrived from China during the 8th century. During the 1200s, Kublai Khan led the Mongols to seize power in China. This prompted further migration to the region, which would later become Laos. During the following century, a Khmer empire was established here. It was known as *Lan Xang* – literally "Million Elephants." The Lao people organised themselves into Lao principalities.

During the 19th century, France persuaded one of the leaders to accept a French protectorate, as insurance against conquest by China or Siam. France then united all of the Lao principalities into one country. Thus came to be the name Laos, which is the plural of Lao.

France used Laos as a buffer between their Asian acquisitions and British Burma. The French imposed a Vietnamese-staffed civil service in Laos, but they did not contribute to the protectorate. When the Japanese occupied French Indochina in 1941, the Lao people obtained more autonomy than they had experienced under French protection. Following WWII, France tried to take back Laos; in 1949 the latter was declared an independent associate state of the French Union. The United Nations recognised Laos as a separate country, and in 1953 France allowed Laos to become a monarchy. In 1954, communists occupied areas of two Lao provinces, namely Phong Saly and Sam Neua. In 1959, King Savang Vatthana became monarch of the Lao kingdom. By 1960, the communist-led *Pathet Lao* (Lao People's Party) forces made advances, and during a three-day period in December that year, Vientiane was ruled by four successive governments (White, 1961a).

In 1962, the Royal Lao Government and the State of Israel embarked on a joint experimental farm, along the Mekong River. Known as the Vientiane Pilot Project, this project provided farmers with seed, fertiliser, insecticide and even irrigation. The objective was to transform subsistence farmers into market-oriented entrepreneurs. Participants prospered and by 1968, 250 families had participated in the project (White, 1968).

In December 1975, the Lao Patriotic Front – the political arm of the *Pathet Lao* – abolished the monarchy and created a communist[31] entity, the Lao People's Democratic Republic. All Christian seminaries were closed that year. Zasloff & Unger (1990) presented a detailed account of Laos beyond the revolution.

---

[31] For a thorough discussion of the communist movement in Laos, see: Brown, Zasloff & Staar (1986).

Traditional Lao Home

French Colonial Mansion in Vientiane

Centrally Planned Policy & Transition in Laos

When the Lao People Revolutionary Party took control of the Lao People's Democratic Republic, in 1975, it implemented a policy of accelerated socialisation. Harsh policies shunned entrepreneurship and co-operatives replaced private initiatives. Former royalists were sent to re-education camps where they were forced to accept communism. In 1977, Laos and Vietnam signed a treaty of friendship and co-operation.

In 1987, Laos implemented its New Economic Mechanism – a radical shift in public policy. This recognised market forces as legitimate, and began liberalising the centralised economy. The first national election took place in 1989. Inflation reached 76% that year. In 1990, the state launched a drive to attract foreign investment, privatise state enterprises, develop import-substitution industries and promote exports. In 1991, a new constitution ushered in further economic reforms. New laws were subsequently introduced, governing property, labour and foreign investment. New relationships were also developed. Laos signed treaties of friendship and co-operation with Thailand in 1991, with Cambodia in 1992, with China in 1994, and with Myanmar in 1994.

In April 1994, Laos and Vietnam signed an agreement on goods in transit, which allowed these commodities to be transported across either Laos or Vietnam, on the way to the other. It was expected that this would facilitate international trade. In May 1994, Laos introduced a liberal law governing investments. This streamlined foreign investment regulations and tax structures. Legislation included tax holidays, a 1% import duty on capital goods associated with production, and a flat-rate corporate tax of 20%. This was to lead to a major influx of foreign capital to create joint ventures, as well as 100% foreign-owned investments in commerce, industry and services. Furthermore, the government committed itself to expedite the business application process. Japanese and Taiwanese investors expressed considerable interest. Until 1996, Laotians needed permission to change residences; then, this requirement was done away with.

On July 23, 1997, Laos joined ASEAN. In 2000, GDP growth was 4.5%. Yet, GDP per capita was a mere $272 that year, and half the nation's budget came from foreign aid. In March 2001, Laotian and Thai officials initialled an agreement to build a second bridge to link the two countries; the geographic position of Laos could help it become an important assembly and trans-shipment centre.

Water-Buffalo in an Agrarian Economy

Local Industry is Limited

The Traditional Belief System

W. Robert Moore, Chief of Foreign Editorial Staff at *National Geographic*, wrote about Vientiane, the capital of Laos, "Biggest buildings in town, except for a few government offices are the Buddhist temples (1954, p. 666)." Even today, religious beliefs are very important in Laos, and Laotians are noticeably influenced by folk tales, superstitions, and traditional animist beliefs – in addition to the ancient beliefs of the national religion, Theravada Buddhism. An article in the August 28, 1992 issue of *Asiaweek* explained that it is commonly believed that the Mekong River gets "hungry" for human souls, without which the annual rains will not arrive:

> a little girl (was) swept away by the current while picnicking with her family on a sandbank. Her mother and father made no attempt to save her. Two foreigners snatched the child from the swift water after a desperate effort. The parents were fearful because the river had been thwarted in claiming the child (p. 63).

Preparing Frogs for Lunch

Across Laos, Theravada monasteries, known as wats, dominate every town, and almost every house, shop and office has a private temple. It is even common for boats cruising the Mekong River, to dock for the crew to jump ashore, light incense and pray. The Lao wats are architecturally distinct from monasteries elsewhere in Asia. In Laos, wats have large terraces, and flare symbols on the roofs.

Religion is Paramount

Theravada Monk

Theravada monks are highly influential in Lao society. They are consulted on virtually all matters, thereby playing a role in a diversity of spheres, ranging from private life to government policy. They have traditionally had a great impact on the educational system; it used to be that the only schools were in wats.

The official calendar used in Laos is that of the Lao Buddhist era – not to be confused with the Thai calendar; the Christian year 2002, for example, corresponds to 2640 in the Lao calendar, and 2545 in the Thai calendar.

Central to the belief system in Laos is the ultimate goal to extinguish unsatisfied desires. Its doctrine focuses on aspects of existence, including *dukkha* (suffering from unsatisfied desire), and *anicca* (impermanence). Assuming that unsatisfied desires cause suffering, suffering can be eliminated if its cause – desire – is eliminated. A respectable person, then, according to this ideology, should not work towards the satisfaction of materialistic desires, but should, rather, strive to eliminate the desire itself. A monk, for instance, is specifically prohibited by the religion, from tilling fields or raising animals.

Lao folk tales reinforce the belief that a male monk should not labour for material wealth; yet, the same folklore conditions women to accept a heavy burden in exchange for honour, protection and security. Even the Lao currency portrays agricultural work being done by women.

Currency Depicts Work in the Fields

Informal Economic Activity

Numerous Lao families who farm during the wet season become self-employed gold-diggers during the dry season. The prospectors camp along the Mekong River, especially in the region of Luang Prabang. The women do the heaviest work, digging for dirt and panning it in wooden trays. The men weigh the gold, up to one gram per day.

The *Far Eastern Economic Review* quoted a London newspaper as saying that "Lao rice farmers have a reputation in this dynamic region for lying down, closing their eyes and listening to their crops grow in fertile paddy fields (1994, p. 60)." Indeed, entrepreneurial spirit is not very prevalent in the traditional Lao belief system. Dana (1995b) addressed this issue in detail.

Pluralism

Laos has a diversity of ethnic minorities with a variety of beliefs, attitudes and traditions. White reported: "There's a Thai Dam...They sacrifice dogs to the spirits. They beat them to death before they roast them, so that the meat will be tender. And here's a Kha Koui. They never wash...(1961a, p. 260)."

According to government sources, ethnic minorities in Laos include: Lan Tene; Meo Khao; Meo Lay; Mouseudam; Mouseudeng; and Mien, formally known as Yao. In addition, Laos is home to eight kinds of Kais, and to seven types of Thais. There are also the Hmong people, referred to by the Chinese as *Miao* – literally meaning barbarians; these are the subject of Garrett (1974).

In all, there are 68 ethnic groups in Laos. White (1987) grouped these into three broad consumer cultures segments: the highlanders, known as Lao Theung; the mountain tribes, called Lao Soung; and the valley dwellers, referred to as Lao Lum. According to United Nations sources, the Lao Lum comprise 60% of the country's population; these people are involved in government, as well as wet-rice agriculture. The Lao Theung people are concentrated in slash-and-burn subsistence agriculture, and they have less contact with consumer culture and the monetised economy. The Lao Soung people are cut off from any legal sector of the monetised economy; some are involved in the smuggling of gems or narcotics.

Given the traditional belief system of Laos, the formal business sector in this country consists largely of people who are not ethnic Laotians. A significant Chinese minority is prominent in small-scale enterprise. This community consists primarily of people – most of whom are these Teochews – who are descendents of migrants from southern China. Vientiane is home to at least 2,500 Chinese entrepreneurs, and these people own a disproportionate percentage of restaurants, movie theatres, hotels, repair shops and jewellery shops in the capital city.

Muslim men are also very active in the small business sector of Laos. Individuals from India and Pakistan are especially active in the garment industry.

In addition, Thai and Australian entrepreneurs have recently created numerous new ventures in Laos. Unlike the Chinese and the Muslims who reside in this country permanently, the Thais and Australians in Laos tend to be sojourners.

Significant Muslim Minority

Entrepreneurship

Whereas American involvement in Lao business tends to be in the form of large business, Thai investments tend to be in the medium-size category, as are Australian ventures. Even more involved in small business are French entrepreneurs.

One problem encountered by entrepreneurs in the manufacturing sector is the poor infrastructure. Given the generally inadequate conditions of the roads – often flooded during the rainy season – and the lack of a railway, a quarter of all traffic in Laos uses the Mekong River. People and buffalo stand side by side, on boats or barges, for hours. Sometimes, cargo gets damaged.

A gift from the people of Australia, the $30 million Friendship Bridge across the Mekong River between Laos and Thailand was opened in 1994. It was hoped that this would increase opportunities for trans-border entrepreneurship.

Have Transport, Will Travel

As for the Lao people, 85% are involved in agricultural sectors. Crops include coffee, corn, rice, tea, tobacco, vegetables and wheat. Given that three fifths of the GNP comes from agricultural output, the government has instituted reforms providing incentives to farmers. The plan introduced preferential tax policies for agriculturalists, and increased investment in the sector, especially in irrigation. The plan also raised the prices of produce, and linked remuneration with output. Although annual per capita rice output in Laos is 350 kilograms, some provinces experience occasional rice shortages. Many farming communities are migrational; they deforest land for a crop, and then move elsewhere. There is a constant breeze of smoke and ashes over the Mekong River. Opium is an important crop, of which Laos produces about 300 tons annually.

Internal Economic Activity

Rice Merchant at the Bazaar

Women Retailers in the Formal Sector

Downtown Vientiane

Toward the Future

Reforms have increased the autonomy of firms, and the state no longer has the monopoly on supply, purchasing and marketing. Yet, O'Driscoll, Holmes & Kirkpatrick (2001) reported that Laos, with its average tariff exceeding 19%, had less economic freedom than any of its neighbours. Furthermore, enterprise is largely limited to women and minorities, as local culture does not encourage it. The local shortage of skilled labour limits manufacturing. Furthermore, the shortage of educated individuals limits the service sector.

Between Women

Chapter 14

The Socialist Republic of Vietnam[32]

Introduction

The Socialist Republic of Vietnam covers 329,560 square kilometres. With shores on the Eastern Sea – also known as the South China Sea – the country borders Cambodia, Laos and China.

Until the 1980s, Vietnam had an economic policy that conformed to the command system. Leaders at the national level made centralised decisions about local production, often without knowledge of local conditions. Produce raised or goods manufactured in one locality were shipped to the central level and then distributed back to the localities, creating huge inefficiencies and losses, due to mould, rats and slippage. Manufacturing was very limited because the French colonialists concentrated on extracting raw materials and emphasised neither industry nor infrastructure. Manufacturing equipment that did exist was old and rusty. In addition, the US-led embargo, that started in 1964, prevented people and firms in Vietnam from legally replacing industrial parts patented in the United States. In late 1986, Vietnam launched *Doi-Moi* (renovation), which laid the path to a free-market economy, personal freedom, and openness to the West.

[32] This chapter includes information obtained from: the Central Bureau of Statistics; the Committee of Foreign Economic Affairs; the Export Development Trading Corporation; the Foreign Trade Development Centre; the Ministry of Commerce; the Ministry of Industry; and the Planning and Trading Department of Artex-Saigon. The chapter also includes material that first appeared in Dana (1994a; 1994b;1994e), and benefited greatly from the personal assistance of Lady Borton, Field Director, American Friends Service Committee, in Hanoi.

Rice is Central to the Economy

Doi-Moi **Ushered in Entrepreneurship as a Complement to Socialism**

Under the command economy, farming had been done through farm co-operatives. These were privatised in 1988, with the land divided among families of the community. The result was a stunning surge in productivity; farmers who had once lolled about on their production teams worked their own fields from "dew to dew." Within a year, Vietnam began to export rice; by 1989, villagers and city people were no longer gaunt. Up to 1990, most trade was with socialist block countries. Until the late 1980s, the state-supported currency differed widely with the real market value, and this led to a parallel rate.

Since Vietnam has never been an industrial country, the government concentrated its efforts on agricultural development, the rural economy and the domestic market. Thus, the great changes of *Doi-Moi* took place first in the agricultural sector.

Reshaping Agriculture

Economic policy in Vietnam has been aimed at creating a market economy. In the attempt to ensure domestic growth, the Socialist Republic of Vietnam implemented a drastic programme of economic reform. Those regulations, which formerly limited the private sector, have been substantially reduced and Vietnam has moved away from being a very highly centralised command economy. By the end of July 1995, Vietnam had joined ASEAN, signed a treaty with the European Union, and established normalised diplomatic relations with the United States.

Historical Overview

Tonkin, the northern part of today's Vietnam, was originally inhabited by different ethnic groups with the Viet predominating. It was conquered by the Chinese in 111 BC and became a province of the Chinese Empire. Further south was the Kingdom of Champa, founded in the 3rd century. It eventually came to be under the control of the Viet peoples, as they expanded from the north. After a lengthy civil war, the Vietnamese kingdom was split into two in 1660, when the Nguyen family established their own separate kingdom. The extreme south of Vietnam, known as Cochin-China, remained under Khmer rule until a century later. Finally, the whole of Vietnam was unified in 1802, by the Nguyen dynasty. Its capital was Hué.

The people of the north are of a Sino-Tibetan race that has strong Han influences, from many years of Chinese domination; the people of the south tend to have been exposed to slightly more Dionysian and fewer Promethean values. While the people of the north adopted a northern sect of Buddhism, those of Cochin-China included followers of a southern sect, with some Hindu influence.

In 1859, in order to protect persecuted Catholic missionaries and their economic interests in the region, France invaded the south. In 1867, Cochin-China became a colony. In 1885, France established a protectorate over Tonkin and to the south. Hence, France controlled all of today's Vietnam, which it grouped with Cambodia and Laos to create a new entity called Indochine française (French Indochina). In 1887, Hanoi became the seat of the French colonial government. Many resented French rule, and Confucian scholars led some of the earliest revolutionary movements. Particularly famous is Phan Boi Chau, whose "Go East Movement" sent young revolutionaries to study in Japan. One candidate decided to go west instead; Nguyen Tat Thanh – later known as Ho Chi Minh (Ho the Enlightened) – left Sai Gon (later Saigon) in 1911 on a French steamer, on which he worked as a cook's assistant. In February 1930, using the name Nguyen Ai Quoc (Nguyen the Patriot), he got together with other revolutionaries in Hong Kong, where they established the Indochinese Communist Party, with the hope of overthrowing the French colonial administration. In the fall of 1940, Japan invaded, but the Japanese kept the French administration in place.

Ho Chi Minh returned to the region in early 1941, and lived in a cave in Pac Bo, in Cao Pang Province, on the Vietnamese side of the Chinese border. In 1941, he drew together competing nationalist factions to form the *Viet Nam*

Doc Loc Dong Minh Hoi (League for the Independence of Viet Nam), known as the Viet Minh. When the Japanese army confiscated rice from Vietnamese peasants, two million Vietnamese people starved to death (Dinh, 1989).

Traditional Homes

The French Brought Their Cars

French Colonial Architecture in Saigon

On August 16, 1945, within days of the Nagasaki bombing, Viet Minh gathered at Tan Trao, in the northern highlands, where they planned the August Revolution; on August 19, they seized key buildings. On August 25, Emperor Bao Dai yielded to the Viet Minh in Hué. After the revolution, Ho Chi Minh entered Hanoi and wrote the *Declaration of Independence* while hidden in the second floor of an import-export company. The declaration was read before 500,000 people in Hanoi, on September 2, 1945. Ho Chi Minh declared himself president of a newly independent country, but his government faced huge challenges. Only 5% of the population was literate. In accordance with the Potsdam Conference Agreements, Nationalist Chinese occupied Hanoi along with French troops.

Less than three weeks after the declaration of independence, the French landed again in the south, taking over Saigon and regaining control of its former colony of Cochin-China. The Associated States of Viet Nam – formerly Tonkin, Annam, and Cochin-China – became a member of the French Union. The spirit of entrepreneurship thrived, and the nation welcomed refugees from communist China. Long described the sidewalks of Saigon:

> Barbers hang mirrors on trees, unfold stools, and are ready for business. Dentists pull teeth and fit gold replacements before admiring audiences. Physicians cup and massage their patients. Herbalists hawk bottled cure-alls, and fortune-tellers feel heads, measure palms, and divine the future. Squatting vendors sell anything from American cigarettes to lottery tickets and incense sticks (1952, p. 289).

Following their defeat at Dien Bien Phu, in 1954, the French were expelled from the region. The cease-fire of July 22, 1954 led to an armistice agreement – in Geneva – that created the Democratic Republic of Vietnam (North Vietnam) north of the Ben Hai River, and the Republic of Vietnam (South Vietnam) to its south. Small-scale land-owners and Catholics fled from the north to the south. Samuels (1955) reported that about forty civilians moved from the south to the north; interviewees suggest that there were many more who went north for regrouping, expecting to return home after elections that were scheduled to take place in 1956. However, according to Eisenhower's memoirs, the United States disallowed the elections, for fear that Ho Chi Minh would win.

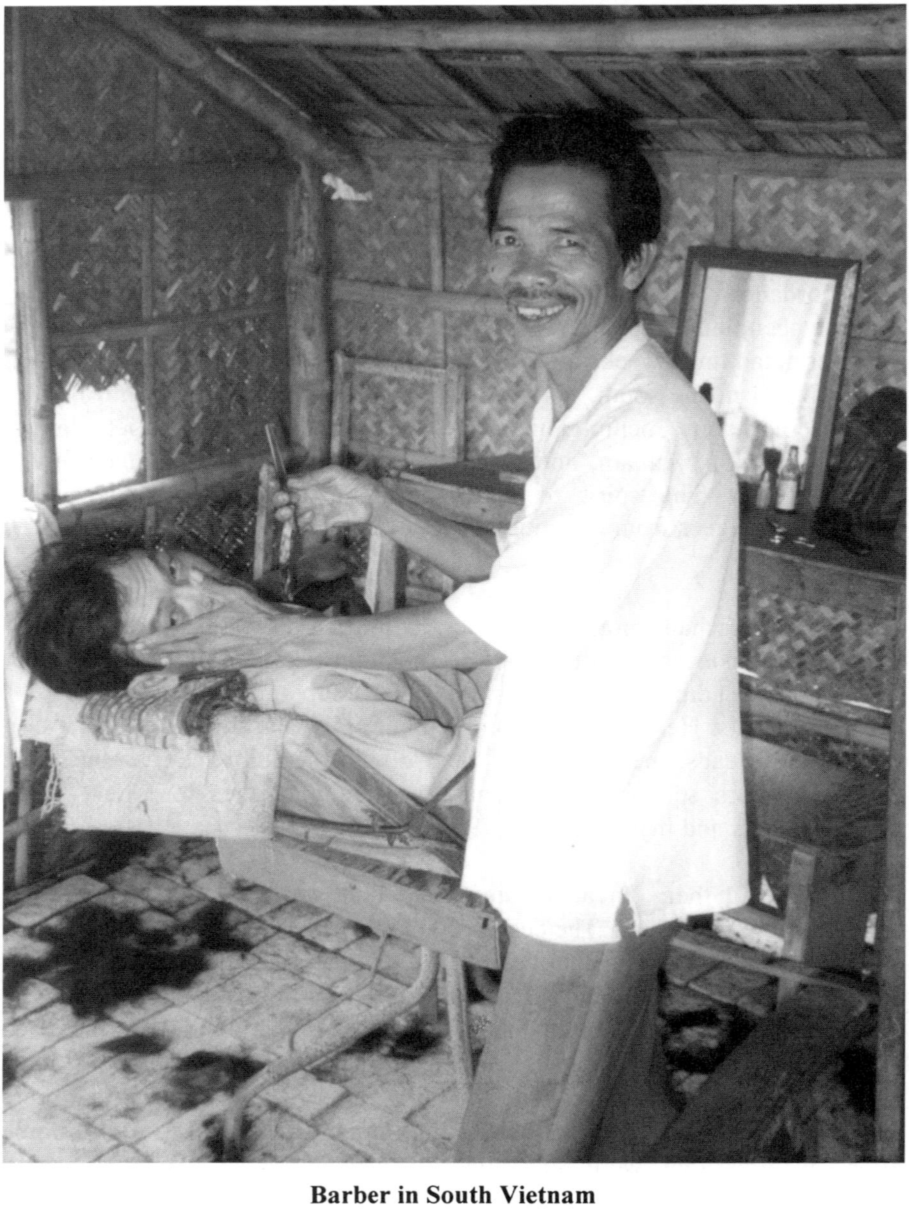

Barber in South Vietnam

According to the Geneva Agreement, the Viet Minh took control of Hanoi on October 10, 1954, and of Hai Duong on October 30. Hanoi, formerly the administrative capital of Indochina, became the capital of North Vietnam. Saigon became the capital of South Vietnam. At the suggestion of Cardinal Spellman, Ngo Dinh Diem – who had been living in the United States for over three years – was brought back to Saigon to lead the government of the South. In North Vietnam, the government – led by Ho Chi Minh – instituted a programme of land reform. The First Five-Year Plan, 1961-1965, put emphasis on basic heavy industries.

Only in South Vietnam could the spirit of entrepreneurship survive – but not for long. The Viet Cong – South Vietnamese who sympathised with the North – fought a civil war, with weapons and leadership from communist North Vietnam and "their inspiration from Peking or Moscow...From dusk to dawn, the Viet Cong ruled nearly half of Viet Nam (White, 1961b, p. 447)." During the war, Sochurek reported, "Despite 30 guards to a train, Viet Cong guerrillas take a fearful toll: Mines and ambushes have killed or injured hundreds of passengers, railroad employees, and soldiers, and have wrecked scores of cars and engines (1964, p. 414)." By 1967, more Americans had died in Vietnam than in the American Revolution of 1776 (White, 1967).

By 1975, South Vietnam, which had formerly been a rice exporter, was living off daily airlifts of rice from Louisiana. This major source of food ceased in April 1975, when the communists took over. The United States extended its embargo to the entire country. The following year saw the creation of the Socialist Republic of Vietnam. With the new regime came a shortage of foreign exchange, resulting in a shortage of imported inputs, new materials and spare parts. This led to under-utilisation of capacity, while domestic supply could not satisfy consumer demand. Further constraints in capacity utilisation were brought about by shortages in energy production and a weak transportation network, which received minimal maintenance and had seen virtually no improvements since the departure of the French. Furthermore, state firms in Vietnam lacked access to modern technology. Their costs of production were high, relative to the quality of products, which tended to be low by international standards. All firms were nationalised. Government newspapers replaced the existing dailies. Private homes were raided and "decadent" literature was burned. Schools of bourgeois learning were closed, as Hanoi spread its Marxist ideology.

Ho Chi Minh

With reunification, all bank accounts in former South Vietnam were frozen, and its people were told they had twelve hours to take their cash to the banks, as the South Vietnamese currency would become valueless. Henceforth, all of Vietnam would have one currency, but each family was limited to a maximum saving, the equivalent of approximately $200.

Chinh gave his rendition of the fourth national congress of the Communist Party of Vietnam:

> The State of the Socialist Republic of Vietnam is a proletarian dictatorship state. On the one hand, it represses counter-revolutionaries, eliminates the comprador capitalist class and the remnants of the feudal landlord class, carries out socialist transformation of the ... private capitalist economic element; at the same time it effects the socialist transformation of the private economic section ... (1977, p. 1).

Socialist Economy

Working for the State

Toward the *Doi-Moi* Model

The Second Five-Year Plan (1976-1980) set a 16-18% targeted annual growth rate for industrial production; actual outcome was 0.6%. In 1978, the Sino-Vietnamese War prompted an exodus of ethnic Chinese from Vietnam. In 1979 and 1980 there were shortages of basic consumer goods including food, as well as shortages of inputs to the industrial sector.

It was liberalisation, in 1981, which helped industrial output to grow an average in excess of 9% during the period of the Third Five-Year Plan (1981-1985). Growth was primarily generated by small and medium-sized enterprises, some in the private sector. The Three-Plan System, introduced in 1981, made this possible:

Plan A: Enterprises operating under this scheme were required to produce using state-supplied inputs and to sell their outputs to the state at low command prices.

Plan B: This scheme permitted firms to acquire inputs on their own, and to sell their outputs independently, provided that the profits were used to purchase additional inputs.

Plan C: This scheme permitted entrepreneurs to diversify and to sell “minor” products with no centrally planned external control.

In 1982, the Fifth Party Congress officially adopted its new economic orientation, recognising the need to: shift emphasis from heavy to light industry; transfer resources to the agricultural sector; and promote exports. As production activities were partially deregulated, individual enterprises were granted some autonomy. In 1984, the government further relaxed restrictions.

When a new currency was introduced, in 1985, old dong were exchanged at a rate of 10:1, but only up to a set quantity. Also in 1985, consumer price subsidies were replaced by wage adjustments. Most prices remained under central control, determined by an average cost-plus formula; however, some costs were not accurately assessed.

The nation’s economic development followed an import substitution model. Industrial policy in Vietnam was influenced by that of the former Soviet Union. Centrally planned industrialisation was aimed at the domestic production of heavy industry such as capital goods. The state supplied the input and capital requirements of enterprises and set quantitative output targets. Policy typically leaned towards achieving self-sufficiency, but neglected opportunities for trade.

Coupled with fighting and the embargo, the command economy led to desperate poverty. The attempt to be self-sufficient had led to insufficient specialisation. Investment had been scattered over too many projects, in a haphazard fashion. Despite vertical integration, there was a lack of horizontal integration. Entrepreneurship was restricted by regulation coupled with excessive bureaucratic centralisation. Innovation and creativity were stifled. Research, science and technology for industry were weak. Energy supplies were lacking. Export-oriented activities were not being given enough attention and there were insufficient links between foreign markets and Vietnamese producers, the latter lacking awareness about international quality, prices and demands. There was more incentive for a farmer to chop down a mango tree and use it for firewood than to harvest the fruit. Aggravating the situation for producers in Vietnam, there were constant limitations on capital resources and raw material supplies, especially imported ones. Construction periods were typically overrun and numerous projects remained unfinished. Reform, or more specifically "renovation" became a necessity.

In Search of Resources

Major change was introduced towards the end of December 1986, when the Sixth Congress of the National Representatives of the Communist Party of Vietnam approved economic reforms that eliminated much of the basic apparatus of control. This programme was called *Doi-Moi.* Individual entrepreneurs obtained the right to get involved in light industry. Profits were defined as the difference between the value of sales and *allowable* costs; enterprises remained liable for taxes on profits. Restrictions on wages were abolished and firms were given the right to recruit as per their needs.

In contrast to the experiences in Eastern Europe and in the former Soviet Union – where a market economy was decreed to have replaced communism without a long transition period – *Doi-Moi* evolved slowly and gradually, ushering in entrepreneurship initially as a *complement* to state enterprise, rather than as a *replacement*. The *Doi-Moi* Model marked the path to: transition to a market economy; openness to the West; openness to overseas Vietnamese (Viet Kieu); and greater personal freedom. In 1988, the state divided huge co-operative farms into private holdings, releasing private initiative.

Free From Central Planning

In July 1988, the Resolution of New Regulation for the Non-State Economic Sector called for making entrepreneurs important components of the national economy. On September 5, 1988, the Council of Ministries adopted Decree 139, regulating the implementation of the Law on Foreign Investment in Vietnam; this set forth procedures relating to foreign investment and the corresponding tax structure. In the case of a joint venture, the foreign party usually contributes capital and technology.

International Technology

A further breakthrough in the reintroduction of market-oriented policies followed in 1989, when the predominantly rural population was granted the right to sell output at market prices; furthermore, this right could be inherited. Whereas Vietnam had been near starvation prior to this reform, the nation soon became the world's third largest rice exporter. Without abandoning Marxist ideals, young intellectuals came to the conclusion that a market economy with a private small business sector and entrepreneurship is the quickest means to attain the benefits sought out by socialism. The result has been a unique blend of socialist and free enterprise policies whereby the entrepreneur is the agent for social change as described by Barth (1963; 1967), but in a socialist state, *i.e.*, the government affirmed its commitment to socialism. New billboards began to promote *Doi-Moi*, while others continued urging workers to follow socialism.

Most prices in Vietnam were released from centralised control, in 1989. Exceptions were electricity, petrol and transport. This policy enabled entrepreneurs and the small business sector to begin setting prices as a function of market forces, while keeping energy and transportation costs artificially low, thus indirectly subsidising entrepreneurship. That same year, the State Committee for Co-operation and Investment was established, as the body responsible for foreign direct investment and for providing guidance to foreign entrepreneurs. Taiwanese entrepreneurs were among the first to arrive in Vietnam *en masse*, and Singaporeans soon followed.

Donated Food Labelled "NOT FOR SALE"

In 1990, the Ministry for External Economic Relations was absorbed into the Ministry of Commerce. The former had already issued about 100 permits to provincial and local enterprises – and a similar amount to central government firms – permitting them to export directly and allowing them to keep a substantial proportion of the foreign exchange earned. Also in 1990, the Council of Non-State Enterprises was established, specifically to promote the interests of the private sector. Very significant was Decree 28, adopted by the National Assembly on June 30, 1990. The new legislation was named the "Law on Amendments and Additions to a Number of Articles of the Law on Foreign Investment in Vietnam."

Towns and cities sprouted into a plethora of street stalls, restaurants and hotels; by late 1990, Hanoi had a reliable telephone system. Motorcycles, taxis and private cars competed for space. Formerly quiet towns became bustling cities.

Relaxation of regulations made it possible for foreign entrepreneurs to hold 100% of the equity of a business in Vietnam. The Law of Foreign Investment was enacted to protect assets of foreign enterprises from nationalisation. Legislation also made new foreign-owned ventures income-tax exempt for the first four years of operations.

As the law stands, today, foreign direct investment in Vietnam cannot be expropriated or requisitioned by administrative procedure. Investors have complained, nevertheless, as legislation regarding foreigners in Vietnam is open to interpretation, and differences in opinion have caused problems in recent years.

As well, foreign entrepreneurs operating in Vietnam were given the right to remit profits abroad as well as to remit payments abroad for the provision of technology, services and loans. In contrast to China, where only foreign exchange profits could be remitted abroad, Vietnam had no foreign exchange restrictions at the time. Therefore, according to Article 86 of the Foreign Investment Law, profits earned in Vietnam could be converted into hard currency at the Bank of Foreign Trade. These funds could then be repatriated, according to Article 87.

With Decision 25-CP, on January 21, 1991, Vietnam established that all enterprises belonging to the socialist state would develop plans directed by the market. Between June 24 and June 27, 1991, the Seventh Congress of the National Representatives of the Communist Party of Vietnam – also known as the Seventh Party Congress – reaffirmed its commitment to *Doi-Moi*, showing support for economic reform, while maintaining political stability was held. On the agenda were concerns such as:

> serious inflation; unstable production; increasing unemployment; wages and salaries below subsistence level ... widespread corruption and other evils ... an erosion of cultural, spiritual and moral values; and declining confidence in the Party and the State (Communist Party of Vietnam, 1991, p. 150).

Sub-contracting for Reebok

Foreign Direct Investment by National Panasonic

Confirming the leading role of the Communist Party of Vietnam, and its adherence to Marxist-Leninism, the Seventh Party Congress renewed the strategy of *Doi-Moi*:

> The overall objectives of this strategy up to the year 2000 are to emerge from crisis, stabilise the socio-economic situation, strive to overcome the condition of poverty and underemployment, improve living conditions for the country to develop more rapidly in the early 21st century. Gross domestic product by the year 2000 will be double of that of 1990 (Communist Party of Vietnam, 1991, p. 157).

In 1991, the Viet Kieu community, numbering 700,000 in the United States alone, contributed approximately $500 million to the Vietnamese economy. In 1992, the Labour Ministry announced a new reform aimed at encouraging new jobs. The minimum wage for employees of foreign firms was *reduced* to $30 monthly. That year, the government began the process of privatising state enterprises. The first to be offered to the private sector was the Legamex garment factory, run by Ms. Nguyen Thi Son in Ho Chi Minh City. This firm exported 1,700,000 jackets to Germany in 1992. The embargo still in effect at the time, entrepreneurs ordered clothing in Hanoi, shipped them to South Korea where South Korean labels were put on them, and re-exported to the United States under their quota.

A tight economic policy succeeded in reducing inflation from over 700% in 1986, to 70% in 1990, 68% in 1991, 18% in 1992 and about 10% in 1993. Per capita GDP in 1992 was $125. In 1994, the United States dropped its trade embargo, and American companies rushed into Vietnam (Dana, 1997b). Per capita income jumped from $190 in 1994, to $275 the following year.

In July 1995, Vietnam became the seventh member of ASEAN. Also in 1995, the National Assembly adopted a new Civil Code, which laid the foundation for a market economy. That same year, import duties of 60% were reduced.

A government resolution, on October 21, 1995, established the Ministry of Planning and Investment. One of 23 ministerial-level offices in Vietnam, it is responsible for decisions, strategy, and plans to develop the national economy.

The Eighth Congress of the Representatives of the Communist Party of Vietnam took place in June 1996. The 1996-2000 plan called for $21 billion, in foreign investment, to complement $21 billion in local investments.

In 1997, Vietnam's exports grew by 24%, while inflation was 4%. Although the Vietnamese dong depreciated by about 20% compared to the dollar, exports grew by only 3.5% between January and October 1998. Inflation approached 10%. The Asian Crisis had hit Vietnam, and this was of concern because 70% of Vietnam's exports depended on neighbours in the region (Rondinelli & Litvack, 1999).

The state's Central Committee convened in October 1998, to outline a new strategy. In contrast to policy that had, in recent years, designated entrepreneurs as economic engines of growth, the Central Committee decided that state-owned enterprises would lead the Vietnamese economy, giving priority to the local market. The state was neither rushed to privatise state firms, nor to reform the almost obsolete banking sector. Instead, the government decided to focus its efforts, during 1999, on agricultural development, the rural economy and the domestic market. In April 2001, the Ninth Congress initiated more reforms.

Bananas for the Domestic Market

En Route to the Market

French Colonial Architecture at the Central Market in Ho Chi Minh City

Infrastructure

In 1989, Hanoi's electrical system still fluctuated between 80 and 360 volts, and there was not yet a reliable domestic telephone system. Most Vietnamese had yet to see a computer or a photocopier. Because of the embargo, the banking system could handle few transfers; funds were routed via Laos. The state-owned banks favoured state-owned firms.

A commercial banking decree, "Decree on Banks, Credit Co-operative and Financial Companies," was promulgated by the Council of Ministers in May 1990. This edict set forth conditions and procedures for establishing commercial banking in Vietnam.

In July 1992, Banque Indosuez of France and the Bangkok Bank of Thailand became the first two foreign banks to open branches in Ho Chi Minh City. By 1993, the ANZ Bank had a branch in Hanoi as well as a representative office in Ho Chi Minh City. Other banks, which were quick to enter Vietnam, included Banque Nationale de Paris, Credit Lyonais, Nordbanke, and the Export-Import Bank of Japan. Thus, the banking infrastructure began taking shape and commercial credit slowly began to evolve. Yet, personal cheques for domestic accounts were still unheard of, and most entrepreneurs paid their employees in cash. With the US dollar worth over 10,000 dong, and 200-dong notes being common, cash transactions were sometimes bulky.

Meanwhile, foreign banks expressed their concerns: they could only repatriate 30% of their capital, and they were not given the three-year tax holiday granted to foreign entrepreneurs. Furthermore, there was a high tax structure for banks: 50% income tax as well as a turnover tax of 4 to 15%.

By 2002, a plentiful and reliable source of electricity made possible many developments. Computer literacy had long surged with personal computers in urban households and neighbourhood Internet cafes offering affordable rates. International organisations, businesses and foreigners used standard bank transfers. Most international organisations transferred salary payments directly into the private bank accounts of staff.

The Impact of the *Doi-Moi* Model

Vietnam's model of economic transition has been stunning; it is quite a unique model, as it has achieved a harmony between government firms still operating under a system of centralisation, and the small business sector operating independently, yet having access to state alternatives as specified by Resolution No. 16. Whereas transition from centralised planning to market economy in the Soviet bloc was very abrupt and economic reform in Yugoslavia very violent (Dana, 1994d; 1994f), the *Doi-Moi* Model of evolutionary change introduced a gradual and smooth transition in Vietnam. Private enterprise is thriving side by side a state-controlled sector.

While many co-operatives have been privatised, most of those still in existence are small to medium-sized businesses. As well, there are countless micro-enterprises in the handicraft and agriculture sectors. The small business sector, including handicrafts as well as small industry, accounts for a significant part of industrial production and exports. Current exports include: bamboo products, bicycle tires and tubes, cotton yarn, glass products, handicrafts, processed forestry products, rubber gloves, and silk yarn.

As a result of *Doi-Moi*, the small business sector in the Socialist Republic of Vietnam now includes co-operatives; family businesses; other private enterprises; and joint ventures between state and private interests. The first three operate free from state control; in the fourth, government influence is limited to contractual agreement.

Although the Socialist Republic of Vietnam has chosen to *retain* socialist ideals, to the ethnographer, Vietnam appears to be thriving more on free enterprise than on Marxist ideology. There is a constant buzz of mercantile energy. Entrepreneurs optimise the use of their minimal resources, and make due despite a poor infrastructure. The spirit of entrepreneurship is in the air, even at the subsistence level. One villa advertises – with a spelling mistake – "Telex Coffee Dansing Massage." Not far, a man sits by the roadside with a pump and fills tires, while another rents out the use of a scale. A coconut stand is never very far away, offering fresh coconut water.

Entrepreneurial Activity in Cai Be

In Ho Chi Minh City, one is overwhelmed by entrepreneurial activity; even children sell a variety of merchandise, including pocket video games, T-shirts, cigarettes, fans, money from French Indochina, and stamp collections. The young peddlers follow their prospective clients, for blocks on end, in an attempt to sell their goods.

Nearby, one shed is a pancake restaurant when it is not being used as a bus depot. "Mr. Fix-It" earns his living by selling a wide variety of inventory, ranging from motorcycles and caviar to tiger-skins for $1,000. Not far, a woman sits on the sidewalk, feathering ducks. A man unloads fresh pork from a wooden container onto the street. Nearby, fish are being laid out in tidy rows; some of the fish are still flopping on the merciless, hot pavement.

At the market, items for sale include bananas, coconuts, dragon fruit, grapes, guavas, mint leaves, oranges, papayas, pineapples, and jeruk – green pomelos sweeter and larger than grapefruit. Chicken, duck, fish, and pork are sold alive or dead, or in between. Also for sale is a mixture of green peas, coconut and cane sugar wrapped in palm leaves fashioned into tiny boxes held together by bamboo toothpicks.

Young Peddler

Self-employed vendors sell food to passengers on state-run trains. While some children run along the aisles fanning passengers for 200 dong, others sell sodas; they open the bottles with their teeth.

Wherever the train stops in the morning, locals have washcloths for rent as well as buckets of water for passengers to refresh themselves. Through the windows of the train, merchants sell bread, coconut water, fruit, juice, pastries, and sodas.

Blind musicians come on board the train to sing or play a melody, while an assistant usually holds out a cap for donations – typically 200 to 500 dong. Simultaneously, one may buy a chicken breast with rice cooked in chicken fat, served on a palm leaf, and eaten with one's fingers. Other passengers prefer to crunch a raw, fertilised duck egg, feathers, bones and all – lightly salted. Still other passengers enjoy having a local cook board the train and flame-broil a fish at their seat.

Informal Business

On the Way to the Station

Selling Through the Train Window

Selling Quail Eggs Aboard the Train

A glance out of the open window supplies passengers with constant stimulus. Catholic churches line the railway tracks side by side with Buddhist temples still in use. Farmers in rice fields, with water buffaloes as tractors, work their land in utter peacefulness while but a few kilometres away, remnants of a bloody war fill the landscape. Graves with crosses, and graves with dragons adorning them dot the landscape, all with the hatred of war still hanging about them.

Some people live a few feet from the railroad tracks in shacks with no kitchen, toilet or running water. Subtle hands reach into the passenger compartment of the train. Others sling a hammock under the railway cars and alight where they choose, while still others enjoy a warm breeze sitting on top of the wagons as clandestine passengers.

In Hoi An, an untouched heritage site, the market is the life of the town. For sale at the market are *chuoi tieu*, a variety of bananas with an attractive colour and taste. A few minutes away, peaceful serenity reigns. Lining the waterfront are storefronts untouched for 200 years or more and buildings inscribed with Chinese characters dating back to the era prior to French rule.

Along the road to Hanoi – in addition to modern petrol stations – stalls display one-litre Coca-Cola bottles filled with kerosene or petrol for sale. Others sell fuel for cigarette lighters.

Every day, 50,000 peddlers transport their goods between Vietnam and China. Micro-enterprises involve exporters piggybacking freight. As porters, women are less expensive than mules; many carry wicker baskets, each balanced at the end of a pole leaning across the shoulders. In the baskets are cats and dogs destined for Chinese dinner tables.

Pluralism

Ever since the French era, the Hoa – a significant ethnic Chinese minority – were very active in the small business sector of Vietnam; the French – not wanting to deal directly with those they had colonised – encouraged the Hoa to serve as "middleman" entrepreneurs in the region, and these continued to dominate sectors of the economy even after 1954. Auster and Aldrich (1984), Bonacich (1973), Cherry (1990), Dana (1997d), and Loewen (1971) discuss middleman entrepreneurship.

The Chinese in Vietnam – of whom 57% are Cantonese – comprise about 3% of this country's population, and they control half the local economy. Cho Lon – the Chinatown of Ho Chi Minh City – was founded by Chinese immigrants in the 18th century, and the settlement was fused into Saigon in 1932 (Long, 1952).

Cyclos weave their way through traffic among motorcycles bound for market, carrying dozens of ducks and chickens. Feathers seem to appear everywhere. On many street corners, food is served, including rice soup, fish, duck, chicken, pork, vegetables and coconut paste. Most dishes are served in re-used plastic bowls, not necessarily washed between users.

Most people in Vietnam are Buddhist, Confucian or Taoist. During colonial rule, the French converted 1,600,000 Vietnamese to Catholicism and many of these fled from the north to the south, in 1954 and 1955 (Samuels, 1955).

Toward the Future

In 1952, George W. Long, writing for *National Geographic*, noted a young lieutenant in Vietnam raising his glass and saying: "To our American visitors. I hope they will return someday. And when they do, I hope they will find this country happy, prosperous, and at peace (Long, 1952, p. 328)."

Half a century later, Vietnam is at peace, its people appear happy, and prosperity is spreading. The Vietnam-US Bilateral Trade Agreement established guarantees against expropriation, and mechanisms for dispute resolution. Yet, O'Driscoll, Holmes & Kirkpatrick (2001), reveal that Vietnam is particularly poor in property rights.

As discussed by Dollar (1999), it would be misleading to think of Vietnam as having state-sector households distinct from private-enterprise households. Instead, it is common for employees of the state to supplement their incomes with self-employment. Although large state enterprises may be inefficient, people in Vietnam demonstrate a strong work ethic. In contrast to China, where individuals dare speak up against their boss, the Vietnamese tend to be very disciplined, and generally have mild temperaments. As state enterprises gradually lose their monopolies, a new middle class of *nouveau-riche* entrepreneurs is emerging.

Path to the Future?

The small business sector will continue to be important to the development of the Vietnamese economy in the future. Firms may be established rapidly, and they can produce quick returns on investment. Given the low level of wages, new ventures are likely to utilise labour-intensive technology, creating considerable employment. Small enterprises are flexible as to location, and may be situated in rural areas, thereby reducing development imbalances.

There are nevertheless constraints. Tan & Lim (1993) reported that 80% of business people surveyed in Vietnam found bureaucracy and corruption to be major obstacles to enterprise. More recently, Venard explained:

> Vietnamese corruption arises from four factors: strict relations between the government and some private interests, decentralisation to regional governments which makes them more powerful and less controllable, under-payment of state employees, and vague laws governing commercial transaction (1998, p. 87).

The government is faced with the challenging task of dismantling bureaucratic structures no longer relevant to the economic model of *Doi-Moi.* In addition, Vietnamese entrepreneurs are faced with a complex tariff scale; local taxes; business turnover tax; and a high personal income tax. Entrepreneurs often prefer to bribe an under-paid tax collector than to pay the taxes that they owe. Whereas Vietnamese entrepreneurs are heavily taxed, foreign direct investment is encouraged. Vietnamese entrepreneurs are therefore getting "foreign devil" companies set up in Hong Kong on their behalf. Profits are made in Hong Kong, resulting in less tax and less bureaucratic interference. A reduction in government intervention would allow entrepreneurs to concentrate more on business than on avoiding government regulation. Furthermore, given that Vietnam still has considerable import duties and that large-scale smuggling is known to occur, the entrepreneur who imports legally finds himself at a disadvantage. Appropriate government action should correct this.

Exports of the Socialist Republic of Vietnam were traditionally destined to former East Germany and other socialist countries where demand exceeded supply. Quality, marketing and advertising were not concerns. Vietnamese entrepreneurs have since learned about marketing, but they fail to understand long-term environmental and sustainability issues. Nor have they come to terms with the fickle nature of capitalism, as their coffee rides the market one year but not the next. Government promotion of entrepreneurship should include the training of entrepreneurs, provision of industrial estates and common facility services, including consultant services and labour relations services. Also beneficial would be access to financing for development, industrial research, and assistance in procuring inputs.

Taiwanese entrepreneurs are important investors in Vietnam. They bring technology, they create jobs, and they allow Vietnam to benefit from their marketing skills and networks abroad, contributing to exports. However, these people are concerned about what they perceive to be excessive government intervention in the economy.

In summary, Vietnam has introduced reforms, but has maintained a hands-on approach. Until recently, considerable resources were being expended, in order to spread the socialist model throughout Vietnam. The

Doi-Moi Model allowed entrepreneurship to gradually play an increasingly important role – without overthrowing the socialist establishment. Institutions and regulatory frameworks are slowly, but continuously, being adapted to the needs of the future.

Waiting for the Future

Chapter 15

Toward the Future

Transition

Transition to a market economy involves profound economic changes, and sometimes – but not necessarily – political change as well. This book has surveyed several models of transition that have been implemented in different parts of Asia (see Table 15.1). Each approach has its advantages and its drawbacks. Success is also influenced by historical experience, cultural values and other factors.

In the PRC, gradual transition, coupled by a strong central government, and an acetic work ethic, has been very successful. In contrast, where rapid transition has undermined traditional institutions, polarisation has often led to social problems, and in some cases to instability. In the Kyrgyz Republic, for example, rapid transition has resulted in a new class of poor, as well as an emerging entrepreneurial class.

While rapid transition has its problems, foreign investors have reason to be concerned when governments are slow to establish legal frameworks for property systems. Writing during the 18th century, Adam Smith emphasised the importance of a stable legal framework (Smith, 1892). This is especially important in transitional economies. Even when governments strive for rapid reform, they can be seriously handicapped by the lack of established legal frameworks for property systems. O'Driscoll, Holmes & Kirkpatrick (2001) found that in the absence of a firm commitment to a solidly established rule of law, even a decline in government intervention has not led to economic freedom. Vague laws provide opportunities for inconsistent discretionary treatment, and this opens the door to corruption.

Recommendations for Policy-makers

During the heyday of the Silk Road, caravans of 50 to 1,000 camels linked the Near East, with the Far East, via Central Asia. In summer, the caravans travelled by night to escape the heat. At risk of being attacked, the richest merchants maintained companies of archers; poorer small-scale merchants paid the rich ones for protection. In many transitional economies today, one also finds poorer small-scale entrepreneurs paying larger organisations for protection. The establishment of a solid legal framework of ownership and the rule of law will be prerequisites for true transition.

A priority for governments should be to determine the appropriate degree of regulation to enact and to enforce, such that the benefits to society exceed the costs of compliance. John Stuart Mill (1869) argued that the only purpose for which power can be rightfully exercised over a member of society, against his will, is to prevent harm onto others. While some regulation is required to ensure order, excessive intervention is counter-productive. Where import duties are considerable, smuggling becomes popular – as is the case in Myanmar and in Vietnam. Indeed, economic transition requires that ideological change be effectively managed. For this, goals must be clear and achievable. It would appear that even a culture supportive of entrepreneurship requires the optimal level of regulation and government intervention.

The optimal level of regulation and government intervention is culture-specific. Policy-makers should therefore keep in mind that the success of a policy or programme in the West does not guarantee equal success elsewhere. For this reason, it is crucial to avoid trans-locating these from one environment to a different one. To be effective, policies and programmes should be appropriate to the culture of a society. As summarised by Lewis, "If a religion lays stress upon material values, upon thrift and productive investment, upon honesty in commercial relations, upon experimentation and risk-bearing…it will be helpful to growth, whereas in so far as it is hostile to these things, it tends to inhibit growth. Where Theravada Buddhism is the backbone of social and cultural values…it may have a restraining effect on the accumulation of wealth and the rise of an entrepreneurial class (1955, p. 105)." Policy-makers should be aware of the cultural attributes of different ethnic groups, and policy should take these differences into account.

Economic development programmes have been introduced in different contexts. Experts have become sceptical of targeted economic development programmes, because these often subsidise the wrong people, with no lasting benefits. It appears that micro-finance programmes will have greater success. In any case, no funding should be distributed in transitional economies without post-loan or post-grant training. Recipients should be familiarised with finance, tax and payroll issues. Otherwise, of what use is capital unless there is knowledge to invest it?

A problem facing many economies in transition is that social equities may limit economic development in the future. The challenge facing these countries is to identify ways to broaden participation in economic development. Priorities should include:

- Reduction of poverty, by accelerating agricultural development, and in some cases by controlling population growth;
- Improvement of training and vocational education;
- Improvement of property rights;
- Improvement of the legal basis for commerce;
- Strengthening financial infrastructure;
- Further liberalisation of trade;
- Revision of policies to attract foreign investment;
- Improvement of the management of government expenditures; and
- Reform of tax policies such as to broaden the state's revenue base, in a fair manner.

Management Implications

If there is one generalisation that can be made about doing business in the transitional economies of Asia, it is that across these vast countries, transactions and profits are a function of networks and relationships. Preferential treatment – when reciprocal – reduce overall transaction costs, thus increasing efficiency, competitiveness and profitability. This is generally true, regardless of the specifics of an environment. Nevertheless, one must again be cautious in attempting to generalise across cultures.

In the West, where the firm-type sector prevails, societies tend to take a form specific to them (Geertz, 1963). They resemble each other in their value system, class structure, family organisation, governance, and economic models. Parsons (1951) identified this over half a century ago. It is even truer today, and we have come to refer to this as globalisation. Modern economies in the Occident are democratic and their mainstream society is secular.

In the transitional economies of Asia, there is substantially less uniformity, than is the case across Western societies. Capitalism in Asia does not necessarily come with democratisation. Political reform is not a co-requisite for economic transition. Some governments choose to have greater control than do others; Turkmenistan's president, at the time of writing, was a former communist dictator and Tajikistan was under military rule.

Likewise, some societies are more religious than others, and there are important differences between religions. Attitudes toward business also vary greatly.

Perhaps the most valuable advice one can give to potential investors is that the importance of cultural differences must never be underestimated. There is neither one Asian model, nor one Asian culture. Just because people live in the same region, it does not mean that they share the same views about business. Likewise, there is not one standard approach to transition in Asia. This is further complicated in pluralistic societies, where unlike cultures each have their own implicit and explicit assumptions. Risk also varies with different types of pluralism. Melting pot pluralism – the situation prevailing when minorities adapt to a secular mainstream society – is stable. In contrast, when ethnic groups do not share a mainstream society, polarisation can result in violence, as has been the case in Myanmar, Tajikistan, and elsewhere. There is no consistency across nations.

Nor is there consistency across time, and managers should keep in mind that in many transitional economies, the newly emerging private sector lacks professional, financial and economic structure. Rules change frequently. What is legal today may be banned tomorrow, and vice versa. Also, the ownership of property is not clearly documented, and the liquidity of immoveable assets is often delayed. Where acquisition is not practical, foreign investors may enter markets via networks. Given that communist planners traditionally emphasised vertical integration, managers might find it necessary to explain how, in a market economy, synergy often comes from horizontal integration.

Western management has mastered the art and science of doing business in the firm-type sector. In the transitional economies of Asia, many opportunities are to be found in the bazaar and in the state-controlled planned sector. In the bazaar, the movement of raw materials, processing, distribution and sales are intertwined activities. The focus on relationships supersedes the products and services that are being exchanged. A sliding price system results in prices that are negotiated, reflecting not only the cost or perceived value of a good or service, but also negotiating skills, the relationship between the buyer and the seller, and possibly the time, as well. Also important is an understanding of the parallel sector.

Informal Vendor

State Store

Bazaar

Implications for Educators

Much education in transitional economies has been aimed at teaching managerial content. This is inadequate. Emphasis should also be placed on values, as well as technical content. In the absence of the values related to sustainable long-term entrepreneurship – such as asceticism, frugality, thrift and work ethic (Weber, 1904-5), managerial skills are not being put to optimal use.

Where privatisation and downsizing of state-owned enterprises cause mass unemployment, there is often a mismatch between market demand and skills available in the workforce. The workforce needs retraining in skills that are in demand. Consequently, the technical content of courses needs to be adapted to changes in the economy.

Due to the lack of employment opportunities, and in the absence of appropriate retraining, many people have become self-employed, often in informal or covert activities; after decades of central planning that considered entrepreneurship to be criminal, the concepts underlying entrepreneurship are not fully understood, and legitimate entrepreneurship is confused with illegal transactions. It would be beneficial, therefore, for educators to promote the acceptance of entrepreneurship.

A Salute to the Soviets

Where a vibrant entrepreneurial class is absent, this may be due to the public policy environment, or to the lack of inherent entrepreneurial characteristics. Where entrepreneurial spirit exists, new venture programmes may further enhance the environment for entrepreneurship, as is the case in the United States. However, in a transitional society with little experience of legitimate entrepreneurship, education should first focus on encouraging an entrepreneurship-friendly ideology.

A problem appears to be that Western experts have been teaching managerial methods, ignoring the fact that people in transitional economies are accustomed to being told exact procedures to follow. Under communism, the mindset was to strive to meet quotas; in a market economy, firms should strive to meet demand. A change in mindset is therefore a prerequisite to internalising managerial skills.

As the economy of a nation becomes increasingly complex, marketing functions will mature and become more specialised. Training will be required to help managers solve new problems of planning, distribution and transportation. A difficulty, however, is that educational initiatives are fragmented. In the Kyrgyz Republic, for example, a variety of external agencies – including the European Union's Technical Assistance to the Commonwealth of Independent States (TACIS), the German Technical Cooperation Agency, the International Labour Organisation and United Nations Development Programme – are doing what they can, but with little, if any, coordination.

Toward Future Research

Transition to a market economy has been prescribed as the means to prosperity. It is important, however, to examine the broader picture. In the Kyrgyz Republic, where transition has taken place relatively quickly, the poor have become poorer, also at a fast pace. While economic growth has been prescribed as the remedy for poverty, experience shows that growth creates problems of its own. Of what good is rapid transition if its adverse effects are uncontrolled? It is useful to look not only at the creation of wealth, but also at its distribution. Myanmar is an example of a rich country with a population that is mostly poor. Class mobility, in such an environment, is often a function of access to bribes rather productive creativity or economic innovation.

It would be a fallacy to attempt to understand economic transition in isolation. The move toward market economics is invariably an element of greater changes taking place in society. From an economic perspective, transition involves an increase in the efficiency of the use of resources, toward the achievement of tangible and measurable objectives. From a human perspective, there are additional issues to consider. One must be weary of transition for the sake of change, *transition should be viewed as the means to a better life for members of society.*

What must change and what need not? The answer depends upon historical context and current situation, as well as the result desired. A variety of models are means to achieve different results. This is where more research is needed.

Classic theories cannot simply be taken and injected into transitional economies, in neglect of the environment in which they are to be placed. Even among formerly Soviet republics, there are important differences. Historical, socio-cultural and economic contexts appear to be important factors affecting transition and the environment for business; societies cannot all adopt legitimate entrepreneurial systems at an equal pace, nor should they be expected to.

Toward the Future

Table 15.1. Contrasting Models (Continued on Next Page)

	CAMBODIA	CHINA	KAZAKHSTAN	KYRGYZ REPUBLIC	LAOS
Model of Transition	Restructuring under constitutional monarchy	Gradual transition, 1 country, 2 systems	Restrucuring (*Perestroika*) by decree	Restrucuring (*Perestroika*) by decree	Centrally planned by communist government
Dominant Ethnic Groups	Khmer: 90%	Han: 92%	Kazakh: 53% Russian: 30%	Kirghiz: 60%	Lowland: 60% Upland: 22%
Male Literacy	48%	90%	99%	99%	70%
Female Literacy	22%	73%	96%	96%	44%
Population (millions)	13	1,285	17	5	6
Fertility (Children/ woman)	4.74	1.82	2.07	3.19	5.12
Population Below Poverty Level	36%	10%	35%	51%	46%
Currency	Riel	Yuan	Tenge	Som	Kip

Going to Work in Laos

Table 15.1. Contrasting Models (Continued)

	Myanmar	Tajikistan	Turkmenistan	Uzbekistan	Viet Nam
Model of Transition	Centrally planned by military	Restrucuring (*Perestroika*) by decree	Reform with limited privatisation	Slow, gradual transition	*Doi-Moi* (partial renovation)
Dominant Ethnic Groups	Burman: 68%	Tajik: 65% Uzbek: 25%	Turkmen: 77%	Uzbek: 80%	Vietnamese: 85%
Male Literacy	89%	99%	99%	99%	97%
Female Literacy	78%	97%	97%	99%	91%
Population (millions)	42	7	5	25	80
Fertility (Children/ woman)	2.3	4.29	3.58	3.06	2.49
Population Below Poverty Level	23%	80%	58%	N/A	37%
Currency	Kyat	Somoni	Manat	Sum	Dong

Which Way to the Future?

Bibliography

Abercrombie, Thomas J. (1964), "Cambodia, Indochina's 'Neutral' Corner," *National Geographic* 126 (4), October, pp. 514-551.

Acs, Zoltan J., & Leo Paul Dana (2001), "Contrasting Two Models of Wealth Redistribution," *Small Business Economics* 16 (2), March, pp. 63-74.

Adas, Michael P. (1974), *The Burma Delta: Economic Development and Local Change on an Asian Rice Frontier 1852-1941,* Madison: University of Wisconsin.

Ahmed, Zafar U., Leo Paul Dana, Syed Aziz Anwar & Peter Beidyuk (1998), "The Environment for Entrepreneurship and International Business in the Ukraine," *Journal of International Business & Entrepreneurship* 6 (2), December, pp. 113-130.

Ahmed, Zafar U., Peter B. Robinson & Leo Paul Dana (2001), "A U.S. Entrepreneur in Moscow," *Entrepreneurship & Innovation* 2 (1), February, pp. 51-58.

Aldrich, Howard E., Trevor P. Jones & David McEvoy (1984), "Ethnic Advantage and Minority Business Development," in Robin Ward & Richard Jenkins, eds., *Ethnic Communities in Business: Strategies for Economic Survival*, Cambridge: Cambridge University Press, pp. 189-210.

Aldrich, Howard E., Ben Rosen & William Woodward (1987), "The Impact of Social Networks on Business Foundings and Profit in a Longitudinal Study," *Frontiers of Entrepreneurship Research*, Wellesley, Massachusetts: Babson College, pp. 154-168.

Aldrich, Howard E. & Roger David Waldinger (1990), "Ethnicity and Entrepreneurship," in W. Scott & J. Blake, eds., *Annual Review of Sociology* 16, pp. 111-135.

Aldrich, Howard E. & Catherine Zimmer (1986), "Entrepreneurship Through Social Networks," in Donald L. Sexton & Raymond W. Smilor, eds., *The Art and Science of Entrepreneurship*, Cambridge, Massachusetts: Ballinger, pp. 3-24.

Anderson, Poul H. (1995), *Collaborative Internationalisation of Small and Medium-Sized Enterprises*, Copenhagen: DJOF.

Arendarski, Andrzej, Tomasz Mroczkowski & James Sood (1994), "A Study of the Redevelopment of Private Enterprise in Poland: Conditions and Policies for Country Growth," *Journal of Small Business Management* 32 (3), July, pp. 40-51.

Arnold, General H. H. (1944), "The Aerial Invasion of Burma," *National Geographic* 86 (2), August, pp. 129-148.

Arnold, Hugh J. & Daniel C. Feldman (1981), "Social Desirability Response Bias in Self-Report Choice Situations," *Academy of Management Journal* 24, pp. 377-385.

Arnold, Hugh J., Daniel C. Feldman & May Purbhoo (1985), "The Role of Social Desirability Response Bias in Turnover Research," *Academy of Management Journal* 28, pp. 955-966.

Auster, Ellen & Howard E. Aldrich (1984), "Small Business Vulnerability, Ethnic Enclaves and Ethnic Enterprise, " in Robin Ward & Richard Jenkins, eds., *Ethnic Communities in Business: Strategies for Economic Survival*, Cambridge: Cambridge University Press, pp. 39-54.

Barth, Frederik, ed., (1963), *The Role of the Entrepreneur in Social Change in Northern Norway*, Bergen: Norwegian Universities' Press.

Barth, Frederik (1966), *Models of Social Organization*, London: Royal Anthropological Institute.

Barth, Frederik (1967a), "Economic Spheres in Darfur," in Raymond Firth, ed., *Themes in Economic Anthropology*, London: Tavistock, pp. 149-174.

Barth, Frederik (1967b), "On the Study of Social Change," *American Anthropologist* 69 (6), December, pp. 661-669.

Barth, Frederik (1981), *Process and Form in Social Life*, London: Routledge.

Beach, William W., & Gerald P. O'Driscoll (2001), "Factors of the Index of Economic Freedom," in Gerald P. O'Driscoll, Kim R. Holmes & Melanie Kirkpatrick, *2001Index of Economic Freedom,* New York: The Wall Street Journal.

Beamish, Paul (1993), "The Characteristics of Joint Ventures in the People's Republic of China," *Journal of International Marketing* 1 (2), pp. 29-48.

Boissevain, Jeremy & Hanneke Grotenbreg (1987), "Ethnic Enterprise in the Netherlands: The Surinamese of Amsterdam," in Robert Goffee & Richard Scase, eds., *Entrepreneurship in Europe: The Social Processes,* London: Croom Helm, pp. 105-130.

Bonacich, Edna (1973), "A Theory of Middleman Minorities, " *American Sociological Review* 38 (5), October, pp. 583-594.

Bonacich, Edna & John Modell (1980), *The Economic Basis of Ethnic Solidarity: Small Business in the Japanese American Community*, Berkeley, California: University of California Press.

Boulnois, L. (1963), *La Route de la Soie*, Paris: Arthaud.

Brenner, Gabrielle A., & Jean-Marie Toulouse (1990), "Business Creation Among the Chinese Immigrants in Montreal," *Journal of Small Business and Entrepreneurship* 7 (4), July, pp. 38-44.

Brown, Josephine A. (1944), "6,000 Miles Over the Roads of Free China," *National Geographic* 85 (3), March, pp. 355-384.

Brown, MacAlister & Joseph Jermiah Zasloff (1999), *Cambodia Confounds the Peacemakers, 1979-1998*, Ithaca: Cornell University Press.

Brown, MacAlister, Joseph Jermiah Zasloff & Richard F. Staar (1986), *Apprentice Revolutionaries: The Communist Movement in Laos, 1930-1985*, Stanford, California: Hoover Institution Press.

Bruton, Garry D. (1998), "Incubators and Small Business Support in Russia," *Journal of Small Business Management* 36 (1), January, pp. 91-94.

Cantillon, Richard (1755), *Essai sur la Nature du Commerce en Général,* London & Paris: R. Gyles; translated into English in 1931, by Henry Higgs, London: MacMillan & Co.

Carsrud, Alan L., Connie Marie Gaglio & Kenneth W. Olm (1986), "Entrepreneurs – Mentors, Networks and Successful New Venture Development: An Exploratory Study, " *Frontiers of Entrepreneurship Research*, Wellesley, Massachusetts: Babson College, pp. 229-243.

Chamard, John & Michael Christie (1996), "Entrepreneurship Education Programs: A Change in Paradigm is Needed," *Entrepreneurship, Innovation, and Change* 5 (3), September, pp. 217-226.

Chandler, David P. (1999), *A History of Cambodia*, Boulder, Colorado: Westview Press.

Chandler, David P., Ben Kiernan & Boua Chanthou (1988), *Pol Pot Plans the Future*, New Haven: Yale.

Chandra, Nayan & Rodney Tasker (1992), "The Gem Stampede," *Far Eastern Economic Review*, July 30, p. 20.

Chantrabot, Ros (1993), *La Republique Khmère (1970-1975)*, Paris: Les Editions L'Harmattan.

Chau, Sandy S. (1995), "The Development of China's Private Entrepreneurship," *Journal of Enterprising Culture* 3 (3), September, pp. 261-270.

Cherry, Robert (1990), "Middleman Minority Theories: Their Implications For Black-Jewish Relations, " *The Journal of Ethnic Studies* 17 (4), Winter, pp. 117-138.

Chinh, Truong (1977), "Firmly Grasp the Proletarian Dictatorship," *Vietnam* 225, September, p. 1.

Chow, K.W. Clement, & Tsang W. K. Eric (1995), "Entrepreneurs in China: Development, Functions and Problems," *International Small Business Journal* 1, pp. 63-77.

Christian, John LeRoy (1942), *Modern Burma: A Survey of Political & Economic Development*, Berkeley, California: University of California Press.

Christian, John LeRoy (1943), "Burma: Where India and China Meet," *National Geographic* 84 (4), October, pp. 489-512.

Clark, Milton J. (1954), "How the Kazakhs Fled to Freedom," *National Geographic* 106 (5), November, pp. 621-644.

Communist Party of Vietnam (1991), *Seventh National Congress Documents,* Hanoi: Vietnam Foreign Languages Publishing House.

Crowne, Doug P., & David Marlowe (1960), "A New Scale of Social Desirability Independent of Psychopathology," *Journal of Consulting Psychology* 24, pp. 349-354.

Cummings, Scott, ed., (1980), *Self-Help in Urban America: Patterns of Minority Business Enterprise*, Port Washington, New York: Kennikat, pp. 33-57.

Dalgic, Tevfik (1998), "Dissemination of Market Orientation in Europe," *International Marketing Review* 15 (1), pp. 45-60.

Dana, Leo Paul (1992), "Entrepreneurship, Innovation and Change in Developing Countries," *Entrepreneurship, Innovation, and Change* 1 (2), June, pp. 231-242.

Dana, Leo Paul (1994a), "A Marxist Mini-Dragon? Entrepreneurship in Today's Vietnam," *Journal of Small Business Management* 32 (2), April, pp. 95-102.

Dana, Leo Paul (1994b), "Economic Reform in the New Vietnam," *Current Affairs* 70 (11), University of Sydney, May, pp. 19-25.

Dana, Leo Paul (1994c), "Entrepreneurship, Innovation and Change in Former East Germany: An Ethnographic Account," *Entrepreneurship, Innovation, and Change* 3 (4), December, pp. 393-401.

Dana, Leo Paul (1994d), "Relever les defis du nouvel environement économique mondial: un ensemble de dix modèles politiques en Afrique, en Amerique, en Asie et en Europe," *Finance & Developpement en Afrique, Revue semestrielle de l'Association des Institutions Africaines du Developpement* I (1), June, pp. 1-4; also published in English under the title "Meeting the Challenges of the New Global Economic Environment: An Overview of Ten Policy Models in Africa, America, Asia and Europe," *Finance & Development in Africa* I (1), June, pp. 1-4.

Dana, Leo Paul (1994e), "The *Doi-Moi* Model: An Ethnographic Account of Entrepreneurship, Innovation and Change in Former French Indo-China," *Entrepreneurship, Innovation, and Change* 3 (1), March, pp. 61-84.

Dana, Leo Paul (1994f), "The Impact of Culture on Entrepreneurship, Innovation, and Change in the Balkans: The Yugopluralist Model," *Entrepreneurship, Innovation, and Change* 3 (2), June, pp. 177-190.

Dana, Leo Paul (1995a), "Entrepreneurship in a Remote Sub-Arctic Community: Nome, Alaska," *Entrepreneurship: Theory and Practice* 20 (1), Fall, pp. 57-72.

Dana, Leo Paul (1995b), "Small Business in a Non-Entrepreneurial Society: The Case of the Lao People's Democratic Republic (Laos)," *Journal of Small Business Management* 33 (3), July, pp. 95-102.

Dana, Leo Paul (1996a), "Albania in the Twilight Zone," *Journal of Small Business Management* 34 (1), January, pp. 64-70.

Dana, Leo Paul (1996b), "Small Business in Mozambique After the War," *Journal of Small Business Management* 34 (4), October, pp. 67-71.
Dana, Leo Paul (1996c), "The Last Days of the Compañero Model in Cuba," *Entrepreneurship, Innovation, and Change* 5 (2), June, pp. 127-146.
Dana, Leo Paul (1997a), "Change, Entrepreneurship and Innovation in the Republic of Kazakhstan," *Entrepreneurship, Innovation, and Change* 6 (2), June, pp. 167-174.
Dana, Leo Paul (1997b), "Pepsi, Vietnam," *Management Case Quarterly* 2 (3), Autumn, pp. 11-13.
Dana, Leo Paul (1997c), "Stalemate in Moldova," *Entrepreneurship, Innovation, and Change* 6 (3), September, pp. 269-277.
Dana, Leo Paul (1997d), "The Origins of Self-Employment," *Canadian Journal of Administrative Sciences/Revue Canadienne des Sciences de l'Administration* 14 (1), April, pp. 99-104.
Dana, Leo Paul (1997e), "Vongpackdy, Laos," *Management Case Quarterly* 2 (3), Autumn, pp. 29-32.
Dana, Leo Paul (1998a), "Models for Entrepreneurship: Towards Cultural Sensitivity," in Rabindra Kanungo, ed., *Entrepreneurship and Innovation: Models for Development,* New Delhi: Sage, pp. 64-77.
Dana, Leo Paul (1998b) "Small Business in Xinjiang," *Asian Journal of Business & Information Systems* 3 (1), Summer, pp. 123-136.
Dana, Leo Paul (1998c), "Waiting For Direction in the Former Yugoslav Republic of Macedonia," *Journal of Small Business Management* 36 (2), April, pp. 62-67.
Dana, Leo Paul (1999a), "Bulgaria at the Crossroads of Entrepreneurship," *Journal of Euromarketing* 8 (4), December, pp. 27-50.
Dana, Leo Paul (1999b), "Entrepreneurship as a Supplement in the People's Republic of China," *Journal of Small Business Management* 37 (3), July, pp. 76-80.
Dana, Leo Paul (1999c), *Entrepreneurship in Pacific Asia: Past, Present & Future*, Singapore, London and Hong Kong: World Scientific.
Dana, Leo Paul (1999d), "Kentucky Fried Chicken in Shanghai," *British Food Journal* 101 (5-6), May, pp. 293-495.
Dana, Leo Paul (1999e), "The Development of Entrepreneurship in Macao and Hong Kong," *Public Administration and Policy* 8 (1), March, pp. 61-72.

Dana, Leo Paul (2000a), "Change & Circumstance in Kyrgyz Markets," *Qualitative Market Research* 3 (2), April, pp. 62-73.

Dana, Leo Paul (2000b), *Economies of the Eastern Mediterranean Region: Economic Miracles in the Making,* Singapore, London and Hong Kong: World Scientific.

Dana, Leo Paul (2000c), "International Management: Culture is of the essence in Asia," Invited Essay in *Financial Times*, Mastering Management Special Section, November 27, pp. 12-13.

Dana, Leo Paul (2000d), "The Hare and the Tortoise of Former Czechoslovakia: Small Business in the Czech and Slovak Republics," *European Business Review* 12 (6), November, pp. 337-343.

Dana, Leo Paul (2001), "Networks, Internationalization & Policy," *Small Business Economics* 16 (2), March, pp. 57-62.

Dana, Leo Paul, Teresa E. Dana & Claudio Vignali (1999), "The Cola Experience," *British Food Journal* 101 (5-6), May, pp. 441-482.

Dana, Leo Paul, Hamid Etemad & Richard W. Wright (2000), "The Global Reach of Symbiotic Networks," *Global Marketing Co-Operation and Networks*, Binghamton: International Business Press, pp. 1-16.

Dandridge, Thomas C., & David M. Flynn (1988), "Entrepreneurship: Environmental Forces Which Are Creating Opportunities in China," *International Small Business Journal* 6 (3), pp. 34-41.

Denzin, Norman K. (1978), *The Research Act: A Theoretical Introduction to Sociological Methods*, New York: McGraw Hill.

Desjardins, Thierry (1997), "Le Cambodge, royaume de la corruption," *Figaro,* January 1, p. 4B.

Dinh, Tran Van (1989), "Hue," *National Geographic* 176 (5), November, pp. 595-603.

Dollar, David (1999), "The Transformation of Vietnam's Economy: Sustaining Growth in the 21st Century," in Jennie I. Litvack & Dennis A. Rondinelli, eds., *Market Reform in Vietnam: Building Institutions for Development*, Westport, Connecticut: Quorum, pp. 31-46.

Dubini, Paola, & Howard E. Aldrich (1991), "Personal and Extended Networks are Central to the Entrepreneurship Process," *Journal of Business Venturing* 6 (5), September, pp. 305-313.

Dunung, Sanjyot P. (1995), *Doing Business in Asia*, New York: Lexington Books.

Dyer, Linda M. & Christopher A. Ross, (2000), "Ethnic Enterprises and Their Clientele," *Journal of Small Business Management* 38 (2), April, pp. 48-66.

Ear, Sophal (1995), "Cambodia's Economic Development in Historical Perspective: 1953-1970," *Berkeley NcNair Journal* 3, Summer, pp. 25-37.

Edwards, Mike, & Gerd Ludwig (1993), "A Broken Empire," *National Geographic* 183 (3), March, pp. 1-53.

Evans, Joel R., & Richard L. Laskin (1994), "The Relationship Marketing Process: A Conceptualization and Application," *Industrial Marketing Management* 23 (5), pp. 432-452.

Fan, Ying, N. Chen & David Kirby (1996), "Chinese Peasant Entrepreneurs: An Examination of Township and Village Enterprises in Rural China," *Journal of Small Business Management* 34 (4), October, pp. 72-76.

Far Eastern Economic Review (1994), "Indochina," May 5, p. 60.

Feige, Edgar L. & Katarina Ott (1999), eds., *Underground Economies in Transition: Unrecorded Activity, Tax, Corruption and Organized Crime*, Aldershot: Ashgate.

Franicevic, Vojmir (1999), "Political Economy of the Unofficial Economy: the State and Regulation," in Edgar L. Feige & Katarina Ott, eds., *Underground Economies in Transition: Unrecorded Activity, Tax, Corruption and Organized Crime*, Aldershot: Ashgate, pp. 117-137.

Furnivall, John S. (1956), *Colonial Policy & Practice: A Comparative Study of Burma and Netherlands India*, New York: New York University Press.

Furnivall, John S. (1957), *A Study of the Social and Economic History of Burma*, Rangoon.

Garrett, W. E. (1971), "Pagan, on the Road to Mandalay," *National Geographic* 139 (3), March, pp. 343-365.

Garrett, W. E. (1974), "The Hmong of Laos," *National Geographic* 145 (1), January, pp. 78-111.

Geertz, Clifford (1963), *Peddlers and Princes: Social Development and Economic Change in Two Indonesian Towns*, Chicago, Illinois: University of Chicago Press.

Glinkina, Svetlana (1999), "Russia's Underground Economy During the Transition," in Edgar L. Feige & Katarina Ott, eds., *Underground*

Economies in Transition: Unrecorded Activity, Tax, Corruption and Organized Crime, Aldershot: Ashgate, pp. 101-116.

Golembiewski, Robert T., & Robert Munzenrider (1975), "Social Desirability as an Intervening Variable in Interpreting OD Effects," *Journal of Applied Behavioral Science* 11, pp. 317-332.

Gomes-Casseres, Benjamin (1997), "Alliance Strategies of Small Firms," *Small Business Economics* 9 (1), pp. 33-44.

Gray, Denis D. (2001), "Working Elephants Make Last Stand in Myanmar," *The Jerusalem Post*, March 13, p. 7.

Gray, Karin & Mark Allison (1997), "Micrcenterprise in a Post-Emergency Environment," *Small Enterprise Development* 8 (4), December, pp. 34-39.

Gronroos, Christian (1989), "Defining Marketing: A Market-Oriented Approach," *European Journal of Marketing* 23 (1), pp. 52-59.

Grossman, Gregory (1977), "The Second Economy of the USSR," *Problems of Communism* 26 (5), pp 25-40.

Hagen, Everett E. (1962), *On the Theory of Social Change: How Economic Growth Begins*, Homewood, Illinois: Dorsey.

Hakansson, Hakan, ed. (1982), *International Marketing and Purchasing of Industrial Goods: An Interaction Approach,* Chicester: John Wiley & Sons.

Hamel, Bernard (1993), *Sihanouk et le Drame Cambodgien*, Paris: L'Harmattan Broché.

Haskell, Martin R., & Lewis Yablonsky (1974), *Crime and Delinquency*, Chicago: Rand McNally.

Henry, Stuart (1978), *The Hidden Economy: The Context and Control of Borderline Crime*, London: Martin Robertson.

Herald Tribune (1988), "U Ne Win is Removed From Office in Burma," July 26, p. 2.

Herbig, Paul A. & Cynthia McCarty (1995), "Lessons to Be Learned from the Soviet and Chinese Socialist Experiments: A Cross-Cultural Comparison of Innovative Capabilities," *Journal of East-West Business* 1 (1), pp. 47-67.

Hirsh, Michael & Ron Moreau (1995), "Risky Business," *Newsweek*, June 19, p. 10.

Hisrich, Robert D., & Guyula Fulop (1995), "Hungarian Entrepreneurs and Their Enterprises," *Journal of Small Business Management* 33 (3), July, pp. 88-94.

Hisrich, Robert D., & Mikhail Gratchev (1993), "The Russian Entrepreneur," *Journal of Business Venturing* 8 (6), pp. 487-497.

Hisrich, Robert D., & Janos Vecsenyi (1990), "Entrepreneurship and the Hungarian Transformation," *Journal of Managerial Psychology* 5 (5), pp. 11-16.

Hla, Lludu U. (1975), *Rice*, Mandalay: Kyipwaye Press.

Hodgson, Bryan (1984), "Time and Again in Burma," *National Geographic* 166 (1), July, pp. 90-121.

Hong, Ma (1990), *Modern China's Economy and Management*, Beijing: Foreign Languages Press.

Huntington, Samuel P. (1996), *The Clash of Civilization and the Remaking of World Order*, New York: Simon & Schuster.

Ivy, Russell L. (1996), "Small Scale Entrepreneurs and Private Sector Development in the Slovak Republic," *Journal of Small Business Management* 34 (4), October, pp. 77-83.

Iyer, Gopalkrishnan & Jon M. Shapiro (1999), "Ethnic Entrepreneurship and Marketing Systems: Implications for the Global Economy," *Journal of International Marketing* 7 (4), pp. 83-110.

Jackson, Karl D., ed., (1989), *Cambodia 1975-1978*, Princeton, New Jersey: Princeton University Press.

Japan Times (1999), "Bear Market," March 7, p. 11.

Jaworski, Bernard J., & Ajay K. Kohli (1993), "Market Orientation: Antecedents and Consequences," *Journal of Marketing* 57, July, pp. 53-70.

Jenkins, Richard (1984), "Ethnicity and the Rise of Capitalism in Ulster," in Robin Ward & Richard Jenkins, eds., *Ethnic Communities in Business: Strategies for Economic Survival*, Cambridge: Cambridge University Press, pp. 57-72.

Johanson, Jan, & Associates (1994), *Internationalization, Relationships and Networks*, Stockholm: Almquist & Wiksell International.

Khup, San (1996), "Macroeconomic Policies with Special Emphasis on Infrastructural Development," in Tan Teck Meng, Low Aik Meng, John

J. Williams, & Ivan P. Polunin, *Business Opportunities in Myanmar*, Singapore: Prentice Hall, pp. 69-79.

Kirzner, Israel M. (1973), *Competition and Entrepreneurship,* Chicago: University of Chicago Press.

Kirzner, Israel M. (1979), *Perception, Opportunity, and Profit*, Chicago: University of Chicago Press.

Kirzner, Israel M. (1982), *Method, Process and Austrian Economics*, Lexington, Massachusetts: Lexington Books.

Kirzner, Israel M. (1985), *Discovery and the Capitalist Process*, Chicago: University of Chicago Press.

Knight, Frank Hyneman (1921), *Risk Uncertainty and Profit*, Boston and New York: Houghton Mifflin.

Kruft, Anton T., & Andrea Sofrova (1997), "The Need for Immediate Support-Structures for Entrepreneurship in Transitional Economies," *Journal of Enterprising Culture* 5 (1), March, pp. 13-26.

Kunstadter, Peter (1972), "Spirits of Change Capture the Karens," *National Geographic* 141 (2), February, pp. 267-285.

Kyi, Khin Maung (1970), "Western Enterprises and Economic Development in Burma," *Journal of Burma Research Society* 53 (1), June, pp. 25-49.

Kyrgyz Republic (1991), *Laws on General Principles of Destatization, Privatization and Entrepreneurship in the Kyrgyz Republic,* Bishkek, December 20.

Lasserre, Philippe & Hellmut Schütte (1995), *Strategies for Asia Pacific*, London: Macmillan.

Lewis, W. Arthur (1955), *The Theory of Economic Growth*, Homewood, Illinois: Richard D. Irwin.

Light, Ivan (1972), *Ethnic Enterprise in America: Business and Welfare among Chinese, Japanese and Blacks*, Berkeley, California: University of California Press.

Light, Ivan (1984), "Immigrant and Ethnic Enterprise in North America," *Ethnic and Racial Studies* 7 (2), pp. 195-216.

Light, Ivan, & Edna Bonacich (1988), *Immigrant Entrepreneurs: Koreans in Los Angeles 1965-1985,* Berkeley, California: University of California Press.

Liuhto, Kari (1996), "The Transformation of the Enterprise Sector in Estonia," *Journal of Enterprising Culture* 4 (3), September, pp. 317-329.

Loewen, James W. (1971), *The Mississippi Chinese: Between Black and White*, Cambridge: Harvard University Press.

Lombardo, Gary A. (1995), "Chinese Entrepreneurs: Strategic Adaptation in a Transitional Economy," *Journal of Enterprising Culture* 3 (3), September, pp. 277- 292.

Long, George W. (1952), "Indochina Faces the Dragon," *National Geographic* 102 (3), September, pp. 287-328.

Long, George W. (1954), "Hong Kong Hangs On," *National Geographic* 105 (2), September, pp. 239-272.

Lowdermilk, Walter C. (1945), "China Fights Erosion with U.S. Aid," *National Geographic* 87 (6), June, pp. 641-680.

Macaulay, R. H. (1934), *History of the Bombay Burmah Trading Corporation, Ltd., 1864-1910,* Colchester, London & Eton: Spottiswoode, Ballantyne & Co.

Martin, James H., & Bruno Grbac (1998), "Small and Large Firms' Marketing Activities as a Response to Economic Privatization," *Journal of Small Business Management* 36 (1), pp. 95-99.

Marx, Karl & Friedrich Engels (1848), *Manifest der Kommunisitischen Partei,* London: J.E. Burghard.

Maung, Mya (1991), *The Burma Road to Poverty*, New York: Praeger.

May, Daw Khin San (1996), "Mandalay: The Handicraft Centre of Myanmar," in Tan Teck Meng, Low Aik Meng, John J. Williams, & Ivan P. Polunin, *Business Opportunities in Myanmar*, Singapore: Prentice Hall, pp. 47-52.

McClelland, David Clarence (1961), *The Achieving Society,* Princeton, New Jersey: D. Van Nostrand.

Mill, John Stuart (1869), *On Liberty,* London: Longman, Roberts & Green.

Min, Pyong Gap (1984), "From White-Collar Occupation to Small Business: Korean Immigrants' Occupational Adjustment," *Sociological Quarterly* 25 (3), Summer, pp. 333-352.

Min, Pyong Gap (1986-7), "Filipino and Korean Immigrants in Small Business: A Comparative Analysis," *Amerasia* 13 (1), Winter, pp. 53-71.

Min, Pyong Gap (1987), "Factors Contributing to Ethnic Business: A Comprehensive Synthesis," *International Journal of Comparative Sociology* 28 (3-4), September-December, pp. 173-193.

Min, Pyong Gap, & Charles Jaret (1985), "Ethnic Business Success: the Case of Korean Small Business in Atlanta," *Sociology and Social Research* 69 (3), April, pp. 412-435.

Moore, W. Robert (1950), "Strife-torn Indochina," *National Geographic* 98 (4), October, pp. 499-510.

Moore, W. Robert (1954), "War and Quiet on the Laos Frontier," *National Geographic* 105 (5), May, pp. 665- 680.

Moore, W. Robert (1963), "Burma: Gentle Neighbor of India and Red China," *National Geographic* 123 (2), February, pp. 153-199.

Moore, W. Robert, & Maynard Owen Williams (1951), "Portrait of Indochina," *National Geographic* 99 (4), April, pp. 461- 490.

Morgan, Gareth, ed. (1983), *Beyond Method, Strategies for Social Research,* Newbury Park, California: Sage.

Noar, Jacob (1985), "Recent Small Business Reforms in Hungary," *Journal of Small Business Management* 23 (1), January, pp. 65-72.

North, Douglass (1990), *Institutions, Insitutional Change, and Economic Performance*, Cambridge: Harvard University Press.

O'Driscoll, Gerald P., Kim R. Holmes & Melanie Kirkpatrick (2001), *2001 Index of Economic Freedom,* New York: The Wall Street Journal.

Orwell, George (1930), *Burmese Days*, New York: Harcourt-Brace Co.

Outram, Frank, & G. E. Fane (1940), "Burma Road, Back Door to China," *National Geographic* 78 (5), November, pp. 629-658.

Overholt, William H. (1993), *The Rise of China: How Economic Reform is Creating a New Superpower*, New York: Norton.

Parsons, Talcott (1951), *The Social System*, Glencoe, Illinois: Free Press.

Parsons, Talcott & Neil Smelzer (1956), *Economy and Society*, Glencoe, Illinois: Free Press.

Pasquero, Jean (1988), "Comparative Research: The Case for Middle-Range Methodologies," *Research in Corporate Social Performance and Policy* 10, pp. 181-209.

Passantino, Joseph E. (1946), "Kunming, Southwestern Gateway to China," *National Geographic* 90 (2), August, pp. 137-168.

Patton, Michael Quinn (1982), "Qualitative Methods and Approaches: What Are They?" in Eileen Kuhns & S. V. Martorana, eds., *Qualitative Methods for Institutional Research*, San Francisco: Jossey-Bass, pp. 3-16.

Patton, Michael Quinn (1987), *How to Use Qualitative Methods in Evaluation*, Newbury Park, California: Sage.

Peng, Michael W. (2000), *Business Strategies in Transition Economies*, Thousand Oaks: Sage.

Peng, Michael W., & Peggy Sue Heath (1996), "The Growth of the Firm in Planned Economies in Transition: Institutions, Organizations, and Strategic Choice," *Academy of Management Review* 21 (2), pp. 492-528.

Perazic, Elizabeth (1960), "Little Laos Next Door to Red China," *National Geographic* 117 (1), January, pp. 46- 69.

Ponchaud, François (1977), *Cambodia: Year Zero*, New York: Holt, Rinehart & Winston.

Portes, Alejandro, & Robert C. Bach (1985), *Latin Journey*, Berkeley, California: University of California Press.

Portes, Alejandro, & Leif Jensen (1987), "What's an Ethnic Enclave? The Case for Conceptual Clarity," *American Sociological Review* 52 (6), December, pp. 768-771.

Portes, Alejandro, & Leif Jensen (1989), "The Enclave and the Entrants: Patterns of Ethnic Enterprise in Miami Before and After Mariel," *American Sociological Review* 54 (6), December, pp. 929-949.

Portes, Alejandro, & Leif Jensen (1992), "Disproving the Enclave Hypothesis," *American Sociological Review* 57 (3), June, pp. 418-420.

Prasso, Sheridan T. (2001), "The Riel Value of Money: How the World's Only Attempt to Abolish Money Has Hindered Cambodia's Economic Development," *Asia Pacific Issues* 49, January, pp. 1- 8.

Prud'Homme, Remy (1969), *L'économie du Cambodge*, Paris: Presses Universitaires de France.

Robinson, Peter B., Zafar U. Ahmed, Leo Paul Dana, Gennady R. Latfullin & Valentina Smirnova (2001), "Towards Entrepreneurship and

Innovation in Russia," *International Journal of Entrepreneurship and Innovation Management* 1 (2), pp. 230-240.

Rondinelli, Dennis A. (1991), "Developing Private Enterprise in the Czech and Slovak Federal Republic," *Columbia Journal of World Business* 26, pp. 26-36.

Rondinelli, Dennis A., & Jennie I. Litvack (1999), "Economic Reform, Social Progress, and Institutional Development: A Framework for Assessing Vietnam's Transition," in Jennie I. Litvack & Dennis A. Rondinelli, eds., *Market Reform in Vietnam: Building Institutions for Development*, Westport, Connecticut: Quorum, pp. 1-30.

Rosenkrantz, Stuart A., Fred Luthans & Harry W. Hennessey (1983), "Role Conflict and Ambiguity Scales: An Evaluation of Psychometric Properties and the Role of Social Desirability Response Bias," *Educational and Psychological Management* 43, pp. 957-970.

Sachs, Jeffrey (1993), *Poland's Jump to the Market Economy*, Cambridge: MIT Press.

Sacks, Paul M. (1993), "Privatization in the Czech Republic," *Columbia Journal of World Business* 28, pp. 188-194.

Samuels, Gertrude (1955), "Passage to Freedom in Viet Nam," *National Geographic* 107 (6), June, pp. 858-874.

Sanders, Jimmy M., & Victor Nee (1987), "Limits of Ethnic Solidarity in the Enclave Economy," *American Sociological Review* 52 (6), December, pp. 745-767.

Schumpeter, Joseph Allois (1912), *Theorie der wirtschaftlichen Entwicklung*, Munich and Leipzig: Dunker und Humblat.

Schumpeter, Joseph Allois (1928), "The Instability of Capitalism," *Economic Journal* 38, November, pp. 361-386.

Schumpeter, Joseph Allois (1934) *The Theory of Economic Development, Cambridge*, Massachusetts: Harvard University Press.

Schumpeter, Joseph Allois (1939), *Business Cycles: A Theoretical, Historical and Statistical Analysis of the Capitalist Process*, New York: McGraw Hill.

Schumpeter, Joseph Allois (1942), *Capitalism, Socialism and Democracy*, New York: Harper & Row.

Schumpeter, Joseph Allois (1947), "The Creative Response in Economic History," *Journal of Economic History* 7, November, pp. 149-159.

Schumpeter, Joseph Allois (1949), "Economic Theory and Entrepreneurial History," in *Change and the Entrepreneur*, Cambridge, Massachusetts: Harvard University Press.

Sealy, Thomas S. (2000), "Kazakhstan," *International Country Risk Guide* XXI (9), September, pp. 77-80.

Seglin, Jeffrey L. (1990), *The McGraw Hill 36-Hour Marketing Course*, New York: McGraw Hill.

Sein, Thaung (1950), *Problems of the Burmese Film*, Rangoon: Bamakhit Press.

Shirk, Susan (1993), *The Political Logic of Economic Reform in China*, Berkeley: University of California Press.

Siu, Wai-Sum, & David A. Kirby (1995), "Marketing in Chinese Small Business: Tentative Theory," *Journal of Enterprising Culture* 3 (3), September, pp. 309-342.

Smith, Adam (1892), *Inquiry into the Nature and Causes of the Wealth of Nations*, Vols. I & II (reprinted from the sixth edition with an introduction by Ernest Belfort Bax), Covent Garden, London: George Bell & Sons, York Street.

Sochurek, Howard (1964), "Slow Train Through Viet Nam's War," *National Geographic* 126 (3), September, pp. 412-444.

Speece, Mark, and Phyu Phyu Sann (1998), "Problems and Conflicts in Manufacturing Joint Ventures in Myanmar," *Journal of Euro-Asian Management* 4 (3), December, pp. 19-43.

Steinberg, David I. (1982), *Burma: A Socialist Nation of Southeast Asia*, Boulder, Colorado: Westview Press.

Steinberg, David J. (1959), *Cambodia*, New Haven: Human Relations Area Files Press.

Stewart, B. Anthony (1944), "Salt for China's Daily Rice," *National Geographic* 86 (3), September, pp. 329-336.

Štulhofer, Aleksandar (1999), "Between Opportunism and Distrust: Socio-Cultural Aspects of the Underground Economy in Croatia," in Edgar L. Feige & Katarina Ott, eds., *Underground Economies in Transition: Unrecorded Activity, Tax, Corruption and Organized Crime*, Aldershot: Ashgate, pp. 43-63.

Swe, Myint (1972), *History of Burmese Printed Books*, Rangoon: Kayathuka Press.

Swerdlow, Joel L. (1995), "Burma: The Richest of Poor Countries," *National Geographic* 188 (1), July, pp. 70-97.

Tan, C. L. & T. S. Lim (1993), *Vietnam: Business & Investment Opportunities,* Singapore: Cassia.

Thein, Myat (1996), "Socio-economic and Cultural Background of Myanmar," in Tan Teck Meng, Low Aik Meng, John J. Williams, & Ivan P. Polunin, *Business Opportunities in Myanmar*, Singapore: Prentice Hall, pp. 17-24.

Thomas, Kenneth W., & Ralph H. Kilmann (1975), "The Social Desirability Variable in Organizational Research: An Alternative Explanation for Reported Findings," *Academy of Management Journal* 18, pp. 741-752.

Venard, Bernard (1998), "Vietnam in Mutation: Will It Be the Next Tiger or a Future Jaguar?" *Asia Pacific Journal of Management* 15, pp. 77-99.

Wai, U Tun (1955), *Burma's Currency and Credit*, Calcutta: Orient Longmans.

Waldinger, Roger David (1984), "Immigrant Enterprise in the New York Garment Industry," *Social Problems* 32 (1), October, pp. 60-71.

Waldinger, Roger David (1986a), "Immigrant Enterprise: A Critique and Reformulation," *Theory and Society* 15 (1-2), pp. 249-285.

Waldinger, Roger David (1986b), *Through the Eye of the Needle: Immigrants and Enterprise in New York's Garment Trades*, New York: New York University Press.

Waldinger, Roger David, & Howard E. Aldrich (1990), "Trends in Ethnic Business in the United States," in Roger David Waldinger, Howard E. Aldrich, Robin Ward & Associates, *Ethnic Entrepreneurs: Immigrant Business in Industrial Societies*, Newbury Park, California: Sage, pp. 49-78.

Waldinger, Roger David, Howard E. Aldrich, & Robin Ward (1990), "Opportunities, Group Characteristics and Strategies," in Roger David Waldinger, Howard E. Aldrich, Robin Ward & Associates, *Ethnic Entrepreneurs: Immigrant Business in Industrial Societies*, Newbury Park, California: Sage, pp. 13-48.

Waldinger, Roger David, David McEvoy, & Howard E. Aldrich (1990), "Spatial Dimensions of Opportunity Structures," in Roger David Waldinger, Howard E. Aldrich, Robin Ward & Associates, *Ethnic*

Entrepreneurs: Immigrant Business in Industrial Societies, Newbury Park, California: Sage, pp. 106-130.

Ward, Robin (1987), "Ethnic Entrepreneurs in Britain and in Europe," in Robert Goffee & Richard Scase, eds., *Entrepreneurship in Europe: The Social Processes*, London: Croom Helm, pp. 83-104.

Ward, Robin, & Richard Jenkins, eds. (1984), *Ethnic Communities in Business: Strategies for Economic Survival*, Cambridge: Cambridge University Press, pp. 105-124.

Weber, Max (1904-5), "Die protestantische Ethik und der Geist des Kapitalismus," *Archiv fur Sozialwissenschaft und Sozialpolitik* (20-21); translated (1930) by Talcott Parsons, *The Protestant Ethic and the Spirit of Capitalism*, New York: George Allen & Unwin.

Weber, Max (1924), *The Theory of Social and Economic Organization*, New York: The Free Press.

Webster, Frederick E. (1992), "The Changing Role of Marketing in the Corporation," *Journal of Marketing* 53, October, pp. 1-17.

Wei, Liu (2001), "Incentive Systems for Technical Change: The Chinese System in Transition," *International Journal of Entrepreneurship & Innovation Management* 1 (2), pp. 157-177.

White, Peter T. (1961a), "Report on Laos," *National Geographic* 120 (2), August, pp. 241-275.

White, Peter T. (1961b), "South Viet Nam Fights the Red Tide," *National Geographic* 120 (4), October, pp. 445-489.

White, Peter T. (1967), "Behind the Headlines in Viet Nam," *National Geographic* 131 (2), February, pp. 149- 193.

White, Peter T. (1968), "The Mekong River of Terror and Hope," *National Geographic* 134 (6), December, pp. 737-787.

White, Peter T. (1982), "Kampuchea Wakens From a Nightmare," *National Geographic* 161 (5), May, pp. 590-623.

White, Peter T. (1987), "Laos Today," *National Geographic* 171 (6), June, pp. 772-795.

Williams, E. E., & Jing Li (1993), "Rural Entrepreneurship in the People's Republic of China," *Entrepreneurship, Innovation, and Change* 2 (1), pp. 41-54.

Williamson, Oliver E. (1985), *The Economic Institutions of Capitalism*, New York: Free Press.

Williamson, Oliver E. (1996), *The Mechanisms of Governance*, New York: Oxford University Press.

Wong, Bernard (1987), "The Role of Ethnicity in Enclave Enterprises: A Study of the Chinese Garment Factories in New York City," *Human Organization* 66 (2), Summer, pp. 120-130.

Wu, Yuan Li (1983), "The Role of Alien Entrepreneurs in Economic Development: An Entrepreneurial Problem," *American Economic Review* 73 (2), May, pp. 112-117.

Young, Frank W. (1971), "A Macrosocialized Interpretation of Entrepreneurship," in Peter Kilby, ed., *Entrepreneurship and Economic Development,* New York: The Free Press, pp. 139-150.

Zapalska, Alina M. (1997), "A Profile of Woman Entrepreneurs and Enterprises in Poland," *Journal of Small Business Management* 35 (4), October, pp. 76-82.

Zasloff, Joseph Jermiah, ed., (1988), *Postwar Indochina: Old Enemies and New Allies*, Washington, DC: US Government Printing Office.

Zasloff, Joseph Jermiah, & Leonard Unger (1990), *Laos: Beyond the Revolution*, New York: St. Martin's Press.

Zerbe, Wilfred J., & Delroy L. Paulhus (1987), "Socially Desirable Responding in Organizational Behavior: A Recognition," *Academy of Management Review* 12 (2), pp. 250-264.

Zhuplev, Anatoly V., Fred Kiesner, Asyleeck B. Kozhakmetov, Tan Wee Liang & Alexander Konkov (1998), "Traits of Successful Business Owners: A Comparative Study of Entrepreneurs in Singapore, the USA, Russia and Kazakhstan," *Journal of Enterprising Culture* 6 (3), pp. 257-268.

Zineldin, Mosad Amin (1998), "Towards an Ecological Collaborative Relationship Management," *European Journal of Marketing* 32 (11-12), pp. 1138-1164.

Index

C

D

E

F

G

H

I

J

N

T

U